The War of
American Independence
1775–1783

The War of American Independence 1775–1783

Stephen Conway
University College London

Edward Arnold
A member of the Hodder Headline Group
LONDON NEW YORK MELBOURNE AUCKLAND

First published in Great Britain in 1995 by
Edward Arnold, a division of Hodder Headline PLC,
338 Euston Road, London NW1 3BH

Distributed exclusively in the USA by
St. Martin's Press Inc.,
175 Fifth Avenue, New York, NY10010

British Library Cataloguing in Publication Data
A catalogue record for this book is available from the British Library

ISBN 0 340 62520 1 (Hb)
ISBN 0 340 57626 X (Pb)

1 2 3 4 5 95 96 97 98 99

Typeset in Times by GreenGate Publishing Services, Tonbridge, Kent
Printed and bound in Great Britain by Mackays of Chatham

Contents

List of Maps

General Editor's Preface

Military history has come out of the closet. At the beginning of this century its academic founding father, Hans Delbrück, tried simultaneously to straddle the scholarly preoccupations of German universities and the policy concerns of the Prussian general staff. His fate seemed to be to satisfy neither. Soldiers resented the puncturing of a cherished inheritance; professors doubted whether the study of war was a proper subject for halls of learning. So dominant and so effective was academic scepticism that, with modifications, Delbrück remained the exception that until the 1960s proved the rule.

Furthermore, scholars were often justified in their doubts about military history. Written, it seemed, by soldiers for soldiers, it too often failed to put military operations in a wider context, and subordinated the events of the past to the didactic needs of the present. But, since the 1960s, all this has changed. Spearheaded in the English language by Gordon Craig, Michael Howard and Peter Paret, military history now embraces a formidably large and increasingly diverse group of scholars. Wars are not just studied for their causes and the consequences but as phenomena worthy of proper consideration in their own right.

It may seem strange that the decade associated with student rebellion, opposition to the Vietnam war, and resistance to conscription should also be identified with the establishment of military history as a subject for academic study. But, in retrospect, one was the reverse of the other. If war remained an important factor in modern society, it behoved universities to understand it rather than simply to condemn it. Furthermore the experience of two world wars could not be easily set aside when the dominant factor in the international relations of the Cold War remained the threat of an all-out nuclear exchange. Military history thrived on the back of strategic studies. The fact that a text published in 1832, Clausewitz's *On War*, remained an essential primer helped cement a relationship between history and current theory that was probably closer than in any other area of political science.

But the effect of the 1960s, and of the academic disciplines which emerged out of them, was also to push the study of war and of armed forces into new directions. War as an instrument of social change, armies as political actors, the social composition of the uniformed services, combat as the precipitant of industrial and technological innovation – all these themes, and others, fell within the purview of what has been dubbed the 'new' military history. The subject has been enormously enriched and strengthened by the diversity of approaches the 'new' military history has come to embrace. It is in large measure due to them that historians can raise their heads above campus parapets, and can say with genuine conviction that military history is an essential ingredient of total history.

The purpose of the *Modern Wars* series is to turn this wheel once more. The 'new' military history carries a danger. It is that military history serves every other purpose than its primary one – that it becomes a tool for sociological analysis, a peg for strategic studies, or a defining point for social historians. In the process the war itself can seem to be forgotten. This series will use the insights of other disciplines and other approaches to discuss war rather than use war as the vehicle for those other sorts of history. It is hard to imagine Stephen Conway's book without the bedrock of social, economic, financial, diplomatic and political history on which it rests. But its core is the fighting itself. Indeed his own extensive archival work, reflected in the many illuminating quotations which he cites, vividly conveys the essence of the combat.

To justify its continued status as a subject worthy of separate study, war must represent some sort of continuum. Many of the questions that have to be asked about its conduct are appropriate across time. Indeed the very concept of a series entitled *Modern Wars* would be redundant if this were not true. But the title also begs questions. What is a modern war? Which was the first modern war?

Military historians have tended to produce two sorts of answers to these problems. The first is that of students of the sixteenth and seventeeth centuries, who, following the thesis of Michael Roberts, have spoken of a military revolution which occurred between 1550 and 1650. Part of this revolution was technological – the development of firepower, the refinement of the sailing ship; part of it was conceptual. The growth in the power of monarchies enabled the creation of standing armies, drilled and disciplined the better to exploit both their numbers and their weaponry. The second set of answers identifies the rise of modern war with the French Revolution. Technology has almost no part to play in this set of arguments. Its focus is political and social. By overthrowing the monarchy, the revolution effected a much greater increase in the power of the state than that possessed even by absolutist kings. Civil rights were traded for civic obligations, and so that the state appropriated to itself the right to conscript. The theoretical size of armies increased accordingly, and, more importantly, the ability to sustain major war over a long period became reality. Soldiers were persuaded that they were part of a nation, and that the latter's needs and interests were no more than their own writ large – that the ambitions of the individual were to be identified with those of the wider community.

Stephen Conway endorses neither of these interpretations. For him, the American War of Independence is the first modern war. He argues that the rebels should not be seen simply as English country gentlemen transposed to Virginia and Pennsylvania, and contending that there should be no taxation without representation. Such an interpretation harks back to the constitutional struggles of the seventeenth century, to the English Civil War and to the 'Glorious Revolution' of 1688 (both of them, incidentally, agents in importing the 'military revolution' to the British Isles). Dr Conway, while acknowledging this aspect, points not backwards, but forwards. He sees the war as an agent of revolution, as galvanising Americans with a popular cause, and as making them anxious to throw off the monarchy rather than to treat with it. Alongside the measured vocabulary of the Enlightenment were ranged the populist rallying cries of democracy. Thus radical ideology, particularly in the hands of the American militia, became a force for political polarisation. The effect was to give the war a savagery and intensity suggestive of the Napoleonic wars rather than of the standard (if overdrawn) image of limited, eighteenth-century warfare. British strategy became divided by the sorts of debate characteristic of current counter-insurgency operations. Was the right response to win the 'hearts and minds' of the population, or was it to terrorise and subdue? The need to requisition, and the failure to restrain individual commanders committed to a more extreme approach, pushed the weight in the second direction. On the rebels' side, George Washington may have reflected a more traditional response, with his preference for professional troops and conventional operations, but practice did not accord with prescription. The Americans used their militia and engaged in guerrilla warfare. The social composition of the American army came to anticipate the new order of revolutionary France, not the old order of monarchical Europe. It recruited from across the spectrum of society; it appointed officers from the ranks; it often promoted according to merit, not birth; and many of its soldiers were motivated by the political ideals embraced by the new republic.

Some may see in this a reworking of the 'Atlantic' thesis – the argument that the French Revolution built on the American, that France's participation in the war created the crisis in French finances which prompted the summoning of the Estates General of 1789, and that the French Assembly's 'Declaration of the Rights of Man' was anticipated by the Declaration of Independence. Many of these connections have been discredited and are not revived here. But Stephen Conway is clearly arguing that what happened in America anticipated what happened in Europe. Furthermore, the prefiguring was not simply a matter of style – not simply because the American War of Independence was a revolutionary war. It was also a matter of scale. The war stretched beyond North America. Its most obvious alternative focus was in the Caribbean. But French, Dutch and Spanish participation ensured that the impact was global: its effects were felt in India and Africa, as well as in South America and Europe. The American War of Independence was thus a world war.

Stephen Conway's arguments will not pass without challenge. But they are doubly appropriate – first because they launch the Modern Wars series with the earliest war that the series will treat, and secondly because he concludes that the defining characteristics of the war are those of modern warfare.

Hew Strachan

Preface

This book attempts to combine a broad view of the War of American Independence with a somewhat unorthodox interpretation. The broad view is rarely adopted by American historians, who tend to concentrate on the Anglo-American aspect of the war; their primary concern, understandably, is with the story of the birth of their country. But this narrow approach distorts the conflict. We need to take proper account of its Caribbean, European, Indian, and even African dimensions. British historians, admittedly, are more inclined to present the wide perspective; not least because the loss of America appears more explicable once one recognizes the strains imposed by a world-wide struggle. But the picture of the war offered here differs significantly from the one that most British historians have given. Piers Mackesey – whose masterly study of the conflict remains unsurpassed – portrays the contest as 'the last great war of the *ancien régime*'.[1] He acknowledges elements of novelty in its American aspect, but maintains that the nature of military conflict did not really change until the French Revolutionary War of 1792–1802. Other historians, especially on this side of the Atlantic, have been disposed to follow Mackesey's lead. Jeremy Black, author of a recent account, sees the American war as blending the 'revolutionary' and the 'traditional', but he emphasises the latter rather than the former. The extent of novelty, he argues, can easily be exaggerated.[2] It would be foolish to deny that there were important continuities; at all times some things change more slowly or more subtly than others. Yet I see the scales tipping in favour of novelty – particularly on the American side, and to a lesser extent on the British side too. In my interpretation, the American war appears not as the last of the old order, but the first of the new.

The book starts with a chapter that attempts to provide an explanation of why the Anglo-American war took place. After this brief examination of the background, the second chapter elaborates on the central theme identified above: the case is made for regarding the American war as exhibiting many of the features usually associated with the 1790s. Chapter 3 provides an assessment of the belligerents, discussing their war aims, weighing their military strengths and weaknesses, and considering their ability to tap their economic resources effectively. The fourth and fifth chapters comprise a narrative of the struggle in America; the fourth covering the period 1775–7 and the fifth taking up the story early in 1778, when the imminence of French intervention radically changed the nature of the conflict. Chapter 6 looks at the war beyond America, where the British found themselves from 1778 pitted against the French, then the Spanish, then the Dutch. The immediate impact of the war on the main contestants is examined in Chapters 7 and 8: Chapter 7 focuses on the effects of the conflict on American politics, society and the economy; Chapter 8 is a comparable survey of the war's contemporary influence on the British Isles. Chapter 9 examines the diplomacy of the conflict, from the outbreak of hostilities to the final signing of the peace treaties. The last chapter looks at the longer-term effects of the American war – on the belligerents and on the conduct of warfare.

While writing this book, I had two different sets of readers in mind. The first comprises those seeking a reasonably concise general account of the American war that assumes little or no prior narrative knowledge. For the sake of such readers I have tried to resist the temptation to annotate elaborately – limiting myself, on the whole, to the identification of sources for quotations – and I have included a section on 'Further Reading' which provides a guide to especially helpful books and articles. The second set of target readers is made up of those who know about the war in broad outline, or even in some detail. They will probably be more concerned with analysis and interpretation than narrative: I hope they will find at least some of the chapters thought-provoking. Those wishing to pursue their own research might appreciate the list of manuscript collections consulted for this study which is included in the bibliography at the end; the footnotes, which give some impression of the range of published primary sources employed, might also be helpful.

My notes rarely cite secondary sources, but the section on 'Further Reading' in the Bibliography provides some indication of the great debt that I owe to the legion of scholars who have published articles or books on aspects of the American war: I wish space had been available to acknowledge many other valuable items.

I am also grateful to the owners and custodians of the manuscript collections examined for this book, particularly Viscount Barrington (Barrington Papers); Mr A.C. Bell-Macdonald (Bell-Macdonald Papers); Dr J.L. Campbell of Canna (Campbell of Inverneill Papers); Lt.-Col. R.M.P. Campbell-Preston (Campbell-Preston Papers); Captain C.T.F. Fagan (Tarleton Papers); Olive, Countess Fitzwilliam's Wentworth Settlement Trustees (Rockingham MSS); Mr S.W. Fraser (Spencer-Stanhope

Papers); the Duke of Grafton (Grafton Papers); Mr F.E. Hart (Hart Papers); the Rt. Hon. Baron Home of the Hirsel (Douglas-Home Papers); Baron Hotham (Hotham Papers); the Rt. Hon. Lady Lucas (Polwarth and Robinson Papers); Mr A.R. Maitland (Maitland Papers); the Mellish Trustees (Mellish of Hodsock MSS); the Duke of Northumberland (Percy Papers); Mr Oliver Russell (Macpherson Grant of Ballindalloch Papers); Mrs P.G.C. Somervell (Fettercairn Papers); the Earl of Verulam (Verulam MSS); and the Marquis of Zetland (Zetland (Dundas) MSS). Their co-operation, kindness, and in some cases even hospitality, is greatly appreciated.

I owe much, as well, to the governing bodies of the institutions listed in the 'Manuscripts' section of the bibliography for permission to consult and quote material in their charge. The patience and helpfulness of the archivists and librarians with whom I have come into contact during my research is also much appreciated. Their expertise has enabled me to base this book on a much larger number of documentary sources than I had originally planned.

Scholarship is a collaborative enterprise in another sense. Every author draws inspiration, comfort and encouragement from others. I owe a great debt to my family, for their support, understanding, and patience, and to a veritable army of scholars and friends. Here I must mention David French, who suggested that I write this book and kindly read it in typescript; Peter Marshall and Jeremy Black, who also cast a critical eye over my efforts; Jane Haville, who helped me with material from the Archives Nationales in Paris; Roy Mantel, whose knowledge of Dutch was of great assistance; Ian Christie, my doctoral supervisor and a constant source of encouragement; and Fred Rosen, who allowed me a break from my editorial labours on the letters of Jeremy Bentham (another research interest of mine). I should like also to thank the following for their assistance in a myriad of ways over the years: Martin Daunton, the late John Dinwiddy, John Shy, Michael McGiffert, Ira D. Gruber, Jo Torsella, the late Janet Semple, Philip Schofield, Paul Kelly, Cyprian Blamires, Stephen Lawson, Peter Nockles, Ian McBride, Wendy Davies, John North, Jon Morris, Rick Halpern, Julian Hoppit, Tony Hayter, Bill Speck, the late John Woods, David Murdoch, John Childs, Mark Odintz, Rory Cornish, and Julie Flavell. Those errors of fact or interpretation that remain are, of course, my responsibility alone.

Stephen Conway
University College London

Preface notes

1. Piers Mackesy, *The War for America 1775–1783* (London, 1964), p.4.
2. Jeremy Black, *War for America: The Fight for Independence 1775–1783* (Stroud, 1991), p.58.

Abbreviations

To avoid repetition of details, the following abbreviations have been used for sources frequently cited in footnotes:

Add. MS	Additional Manuscript, British Library
ADM	Admiralty Papers, Public Record Office
American Campaigns	*The American Campaigns of Rochambeau's Army*, trans. and ed. Howard C. Rice Jr. and Anne S.K. Brown (2 vols., Princeton, N.J., and Providence, R.I., 1972)
CO	Colonial Office Papers, Public Record Office
DAR	*Documents of the American Revolution*, ed. K.G. Davies (21 vols., Shannon, 1972–81)
Franklin Papers	*The Papers of Benjamin Franklin*, ed. Leonard W. Labaree *et al.* (29 vols. to date, New Haven, Conn., 1959–)
Greene Papers	*The Papers of Nathanael Greene*, ed. R.K. Showman *et al.* (6 vols. to date, Chapel Hill, N.C., 1976–)
Hamilton Papers	*The Papers of Alexander Hamilton*, ed. Harold C. Syrett and Jacob E. Cooke (26 vols., New York, 1961–79)
HMC	Historical Manuscripts Commission
IOL	India Office Library and Records
Jefferson Papers	*The Papers of Thomas Jefferson*, ed. Julian P. Boyd *et al.* (21 vols. to date, Princeton, N.J., 1950–)

PRO	Public Record Office
RO	Record Office
Sandwich Papers	*The Private Papers of John, Earl of Sandwich,* ed. G.R. Barnes and J.H.Owen (4 vols., London, 1932–8)
SRO	Scottish Record Office
WO	War Office Papers, Public Record Office
Writings of Washington	*The Writings of George Washington,* ed. J.C. Fitzpatrick (39 vols., Washington, D.C., 1931–44)

1

The Anglo-American Crisis

The War of American Independence developed into a conflict that was not simply Anglo-American; but it began as a contest between Britain and her rebellious colonies, and to understand this war within the British empire – without which there would have been no wider struggle – we need to know something of its origins.[1] Arguably, the roots ran very deep; some would say that they can be traced back to the first settlement of English America. Most of the colonies were seventeenth-century creations. Established by royal charters, their systems of government replicated – more or less – that of the mother country. Each colony had a governor, usually the representative and equivalent of the crown; they all, with the exception of Pennsylvania, had a council – in some cases elected but more often appointed – which was like a colonial House of Lords and Privy Council combined; and they all had their own elected assemblies – their own versions of the House of Commons. These English-style governmental structures, together with the remoteness of the colonies from England, led naturally to a sense of local autonomy. Local self-government, however, was often at odds with the intentions of the metropolitan authorities. The potential for conflict was therefore present almost from the beginning.

As early as 1621 an attempt was made to centre the Virginia tobacco trade on England, by forbidding direct sales to continental Europe. In 1637, after an investigation by a special committtee of the Privy Council, Charles I sought to bring Massachusetts under closer royal control. The realization of Charles's aspirations was prevented by domestic problems – war with his Scottish subjects and then civil war in England – but the imperial pretentions of the English Republic were revealed in the Navigation Ordnance of 1651, which was designed to strike at the Dutch stranglehold of the lucrative trans-atlantic carrying trade. With the restoration of the Stuarts in 1660, fresh and more successful attempts were made to assert central control. Navigation Acts, building on the Navigation Ordnance, were passed in 1660, 1663, and 1673. All goods shipped to and from the colonies had to be transported in English vessels, manned predominantly by English crews. Foreigners were squeezed out of colonial commerce and English naval strength was increased. In addition, many colonial exports and nearly all imports were now to be channelled through English ports, thereby increasing customs revenues for the Crown. In 1675 a new instrument of central direction was created, the lords of trade, a permanent committee of the Privy Council which set about overseeing colonial development. It was this body that initiated the disciplining of recalcitrant colonies: persistent defiance of the Navigation Acts by Massachusetts resulted in the loss of its charter in 1684. The policy of metropolitan supervision reached a peak in 1686 with the foundation of James II's Dominion of New England, which was subsequently expanded to take in New York. Local assemblies were abolished and sweeping powers placed in the hands of a royally-appointed governor-general. James's fall saw the collapse of the Dominion and the restoration of assemblies in the northern colonies; but under the new monarchs, William and Mary, efforts to impose closer control of the colonies continued. In 1696 another Navigation Act was passed, and the Board of Trade was formed to oversee imperial commerce.

After 1713, the metropolitan desire for control became less evident. The ending of the long war with France in that year was important because British interest in the American colonies was always more acute in times of war with a European rival. Under Sir Robert Walpole, first minister from 1721 to 1742, a period of peace coincided with limited interest in America. True, the old characterization of Walpole's years as ones of 'salutary neglect' needs qualification. The White Pine Act of 1729 was designed to reserve certain North American trees for the use of the Royal Navy. The Hat Act of 1732 attempted to prevent colonial hatmakers competing with their British counterparts. And, most famously, the Molasses Act of 1733 aimed to reduce American consumption of foreign molasses by imposing duties that discriminated in favour of the British West Indies. But Walpole was loathe to provoke the colonists by rigid enforcement; the Molasses Act in particular was easily circumvented. So long as this laxity continued, all was well. While the British limited their interference in the domestic affairs of the colonies, and made no attempt rigorously to apply legislation affecting their economic activities, the tensions in the imperial relationship were latent rather than manifest.

Renewed warfare disrupted this Walpolian *modus vivendi*. The War of

Austrian Succession convinced British politicians that America – increasingly valuable as an outlet for British manufactures – must be secured against the French. The establishment of a permanent military and naval base at Halifax, Nova Scotia, was the outcome. Even the return of peace brought little reduction in Anglo-French hostility: the treaty of Aix-la-Chapelle of 1748 was regarded by both sides as no more than a truce. It was not surprising, then, that the British continued to show a keen interest. The parliamentary grant in 1748 of a bounty on indigo cultivation in the colonies was an indication of this; another was the act of 1750 that aimed to limit the production in America of finished iron, while encouraging the export to Britain of crude pig and bar iron.

But it was the Seven Years' War that decisively shifted British governmental thinking in a more decidedly imperial direction. From the American perspective, the war was an object lesson in imperial co-operation: locally-raised troops and British regulars had together defeated the French. From the British perspective, however, the conflict revealed the weaknesses inherent in existing imperial arrangements. Only the sending of substantial regular forces to America had brought victory against the French. The British convinced themselves that this was partly because the colonists were such poor soldiers: 'There is no depending upon them in action. …Such rascals ... are rather an encumbrance than any real strength to an army' was the judgement of General James Wolfe.[2] More important was the evidence that the various colonies were incapable of sinking their differences to combat the French threat. The American response had been decidedly patchy: some local legislatures had willingly answered royal requisitions for men or supplies by voting the appropriate money; others, in colonies more remote from the scene of operations, had been less forthcoming. None of this augered well for the future. If the Bourbon powers sought to regain their lost territories, would colonial efforts be any better?

*

Three important decisions flowed from the British experience in the Seven Years' War. The first was to improve the efficiency of existing imperial structures and institutions. The second was to base a British army in America; the conquered provinces needed to be garrisoned and it was thought desirable to keep a body of regular troops available in North America in the event of a renewal of hostilities with the French or Spanish (colonial troops, raised under the requisition system, could be added to this standing army in time of emergency). The third decision was to help to pay for the support of the regulars by taxes raised in America by

The North American Colonies in 1763

parliamentary authority: they had to be financed by parliamentary taxation, or Parliament would lose control of the army, and making the Americans contribute to this would have the clear advantage – from a ministerial point of view – of sparing the hard-pressed British taxpayer. Few British politicians doubted Parliament's competence in this respect. Although Parliament had met only irregularly before 1689, since the Glorious Revolution it had been in session every year and had become an entrenched institution. It had steadily increased its authority, and as early as 1720 claimed the right to legislate for Ireland – even though Ireland had a Parliament of its own. By the 1760s nearly all parliamentarians saw themselves as members of an imperial legislature that could exert its authority in any part of the empire it chose. As the agent of the Connecticut assembly reported from London in March 1765, there was hardly an MP willing to dispute 'ye right of Parliament to tax us'.[3]

George Grenville, the first minister in 1763, started his drive for greater imperial efficiency and an imperial revenue to support the army in America by trying to tighten up the Navigation system. Absentee customs officials were ordered to their posts, and the navy was instructed to act against smugglers. Next came the Revenue or Sugar Act of 1764, which charged new duties on American commerce. This act differed from earlier regulations, however, in that it was not primarily intended to influence the direction of trade, but rather to raise money to pay for the army in the colonies. The Stamp Act, passed in the following year, put duties on legal documents, playing cards and newspapers and was a still more obvious attempt to levy money in America by parliamentary taxation.

The vital ingredient missing from Grenville's imperial vision was colonial consent. When in March 1764 Grenville introduced his budget into the Commons, he let it be known that he was planning to bring in a stamp duty. He added that he would delay the introduction of a Stamp Bill to test American opinion. On 17 May Grenville met the agents of the colonial assemblies and told them that 'his intention by this delay was to have the sense of the Colonies themselves upon the matter, and if they could point out any system or plan as effectual and more easy to them, he was open to every proposition to be made from the Colonies'.[4] It appears, however, that he was already determined to introduce a Stamp Bill and that he was merely expecting the Americans to suggest the proper level or rate of duty, or to argue for certain exemptions. No official communication was sent to the colonies inviting objections or alternative proposals.

Long before the Stamp Act became law, American resistance to parliamentary taxation was all too apparent. At first colonial protest was restrained and moderate. Pamphlets presented the American case, and colonial legislatures sent petitions and memorials to the king and Parliament. Once the Stamp Act was passed, the same methods were pursued on a more extensive scale; but there were new and significant developments as well. Inter-colonial co-operation, so obviously lacking in the Seven Years' War, rapidly emerged in the face of what was seen as the British threat. A Stamp Act Congress, representing nine of the 13 colonies, met in New York and appealed to the British for repeal. Perhaps more alarmingly, from the British standpoint, colonial merchants agreed to boycott British manufactures until Parliament offered relief. Stamp distributors were burnt in effigy, their property was attacked, and they were intimidated into resignation. Colonial governors and other royal officials reported a breakdown of order in some of the major towns. Throughout the colonies shadowy groups known as 'the Sons of Liberty' were formed to co-ordinate local opposition to the Stamp Act; some groups expressed a willingness to go as far as armed resistance if necessary.

The Stamp Act rioters seem to have had other grievances besides parliamentary taxation of the colonies. In New York, Lieutenant-Governor Cadwallader Colden claimed that he was an object of popular displeasure because he had already won the hostility of the province's leading lawyers. Colden's recent insistence on reviewing the evidence in a case in the appellate court, which sat without a jury and usually considered only points of law, had aroused fears of a plot to undermine the age-old right to trial by jury. The lieutenant-governor did not hesitate to accuse his local enemies of fomenting the Stamp Act disorders as a means of settling scores. Similarly, at Newport two of the victims of the Stamp Act riot were already notorious for their part in trying to persuade the London government to call in the Rhode Island charter and bring the colony under closer royal control.

There was also a social dimension to some of the disturbances. In Boston, there were two separate riots. In the first, the stamp distributor Andrew Oliver was the target; in the second, a few days later, the conspicuously wealthy and powerful lieutenant-governor of the colony, Thomas Hutchinson, was the principal victim. According to Francis Bernard, the governor of Massachusetts, popular resentment at the Stamp Act had turned into an attack on property by the lower orders: 'It was now becoming a war of plunder, of general levelling and taking away the distinction of rich and poor so that those gentlemen who had promoted and approved the cruel treatment of Mr. Oliver became now as fearful for themselves as the most loyal person in the town could be'.[5]

Economic grievances were relevant, too. The Seven Years' War had brought prosperity to many of the colonies; when it ended there was a reduction in British government spending on the army and navy in

America and a consequent decline in business for colonial merchants, farmers, and manufacturers. Unemployment was a problem in many of the ports. This post-war depression was exacerbated by British measures. By clamping down on trade with the foreign West Indies, the British government made it more difficult for the Americans to acquire much-needed specie. Simultaneously, the capacity of the colonies to employ alternative forms of money was hit by the Currency Act of 1764, which sought to curtail the printing of paper money to protect the interests of British creditors.

Even so, the main complaint against parliamentary taxation of the colonies was constitutional. The Americans protested that their economic circumstances made taxation grievous; but at the same time they proclaimed their willingness to contribute to imperial defence in the traditional way – that is, through royal requisitions on their own legislatures. What they could not accept was parliamentary taxation. As Englishmen living in America, they believed they shared with Englishmen living in England the right to be taxed only with their own consent or that of their representatives. This was a right that Americans saw as enshrined in the English constitution: it was articulated by Magna Carta, the Petition of Right and the Bill of Rights, and had been secured to the colonists by the charters which provided the basis for their governments.

British parliamentarians saw the matter in a different light. The growing volume of local improvement acts passed by Parliament in the eighteenth century testifies to the confidence in its ability to cater for the interests even of communities like Birmingham and Manchester that were not formally represented in the House of Commons.[6] So, when Americans protested that they were unrepresented in the British Parliament, government writers naturally replied that colonial interests were represented in the same way as were those of Birmingham and Manchester: there were MPs willing to speak for America, just as there were MPs who could voice the concerns of apparently unrepresented towns in Britain. The colonists, wrote Thomas Whately, were 'in exactly the same Situation' as the British non-electors and great corporations like the East India Company: 'none are actually, all are virtually represented in Parliament'. The House of Commons, he went on, represented the interests not of certain communities and individuals but of the whole 'Commons of *Great Britain*' – and that included the Americans, who, after all, were British subjects.[7]

The colonists were unconvinced. There was not, Daniel Dulany of Maryland argued, the same community of interest between unrepresented Americans and British electors and MPs as there was between British non-electors, electors and MPs. Dulany and other colonial writers maintained that the 3,000 miles of ocean separating Britain and the colonies made it impossible for British parliamentarians – even if well-intentioned – to know local circumstances and the ability of the people to bear taxation. Direct American representation in the British Parliament was thought impracticable by most Americans for the same reason: if Americans became MPs in London they would lose contact with their constituents in the colonies. A handful of American MPs, some colonists suggested, would be worse than none: their presence at Westminster would give Parliament the excuse to levy higher and higher taxes on the colonies, until the bulk of British taxation had been shifted onto American shoulders.

The only proper way to raise money in America, the colonists maintained, was through the colonial assemblies – the American equivalents of Parliament. To raise money by parliamentary taxation was to deny Americans control of their own property. Once they had lost that control, they ceased to be freemen and became in effect slaves. As Stephen Hopkins of Rhode Island put it, 'they who are taxed at pleasure by others cannot possibly have any property, can have nothing to be called their own. They who have no property can have no freedom, but are indeed reduced to the most abject slavery'.[8]

The language of freedom and slavery came readily to American lips because they had been brought up to be good Whigs. Their Whiggism, however, was not the Whiggism prevalent in mid-eighteenth-century Britain. By the 1760s, British Whiggism had become so all-embracing as to be almost drained of meaning. American Whiggism, by contrast, was still very much the unreconstructed 'country'-style Whiggism of the first English Whigs, who had come to prominence when England (and the colonies) seemed to be sliding towards despotism under the later Stuarts, Charles II and his Catholic successor James. Colonial Whiggism in the 1760s, then, was still concerned with the old 'country' issues of resisting arbitrary power and defending the integrity of representative institutions; a concern fortified by the frequently-reprinted arguments of the eighteenth-century British 'true', 'real', or 'patriot' Whigs who attacked the corrupt and unprincipled conduct of nominally Whig ministers like Walpole. Many colonists were therefore convinced of the need to guard against any attempt to expand executive power by stealth. In the appropriate circumstances, this cast of mind could easily lead to a firm belief in a plot by members of the executive to acquire unchecked control, or, in the idiom of the day, to destroy liberty and establish tyranny.

The early years of George III's reign seemed to provide ample cause for alarm. Stuart-style despotism appeared to be making a come-back. Not the least sign of this was the position of influence attained by

George's old tutor, the Earl of Bute. Bute had been first minister in 1762–3, and both in Britain and the colonies he was popularly supposed to be controlling policy long after he left office. He suffered from two substantial drawbacks: the country of his birth and his name. The Scots were regarded by many Americans (and Englishmen) as naturally despotic: 'bread up under strange feudal tenets', wrote Landon Carter of Virginia, 'they were strangers to Liberty themselves and wanted the rest of Mankind to live under the same slavish notions that they had ever done'.[9] That Bute's family name was Stuart added colour to suspicions about his intentions. In August 1765, when the stamp distributor and ministerial apologists were hanged in effigy at Newport, Rhode Island, one was labelled as a Jacobite – a supporter of the restoration of the house of Stuart – and another was depicted as receiving a letter from Bute. A print circulating in Boston in November 1765 similarly portrayed the Stamp Act as a product of 'Sc[otc]h Government', and included a symbolic representation of Bute as a jackboot – a common means of signifying his authoritarian predilections by a play on words (his Christian name was John, and Bute could easily be pronounced as boot).[10]

The assault on liberty that Bute was believed to be inspiring or directing was seen as insidiously wide-ranging. For those who wanted to adopt such an interpretation, there appeared to be evidence in abundance. Discussion in Britain of the advantages of establishing an Anglican episcopacy in America particularly horrified New England congregationalists and presbyterians from the middle colonies, and was taken as proof that the parliamentary enactments of 1764 and 1765 were merely the first fruits of a conspiracy to enslave the colonies. Ecclesiastical hierarchies and parliamentary taxes seemed to be part of the same authoritarian design. At bottom, fear of a slide into servitude underpinned American protests; and this fear accounts for their apparently exaggerated response to parliamentary measures that most Britons regarded as moderate and reasonable.

Grenville had fallen from power by the time news of American resistance to the Stamp Act reached Britain. The new ministry, headed by the Marquis of Rockingham but strongly influenced by the king's uncle the Duke of Cumberland, was not initially particularly disposed to meet the colonial demands for repeal. At a cabinet meeting on 30 August 1765 the new ministers agreed that the governor of Virginia, whose assembly had produced the first criticism of British taxation, should be instructed to support 'by all prudent measures the Authority of the British Parliament'.[11] Indeed, while Cumberland lived the ministry was preparing to reimpose order by military means. Instructions to this effect were sent in September and October 1765 to colonial governors and army and naval commanders in America by General Henry Seymour Conway, the administration's secretary of state for the southern department. But reimposing order was not the same as enforcing the act and, after Cumberland's death at the end of October, the ministry started to move tentatively towards repeal. Progress was necessarily slow because, as Rockingham remarked, it was 'no very easy matter to reconcile' the differing views within the ministry.[12] But eventually most of Rockingham's colleagues accepted the need for repeal. It was not American argument that brought this about. The collapse of authority in many of the colonies played a part, and Rockingham appears to have been influenced above all by the fears of the mercantile and manufacturing communities in Britain that American non-importation agreements would have a disastrous effect on the British economy. He used these fears to persuade ministers and MPs that repeal was vital for British interests. He sugared the pill with a Declaratory Act, based on the Irish Declaratory Act of 1720, which proclaimed Parliament's right to legislate for the colonies 'in all cases whatsoever'.

It would be a mistake to assume that Rockingham's view of the empire was radically different from Grenville's. Most of Rockingham's supporters had not objected to the introduction of the Stamp Act, and they repealed it for reasons of expediency. They may have been more prepared than Grenville to listen to the Americans; Rockingham talked of the need to offer the colonists commercial concessions, and the ministry's Free Ports Act of 1766 was designed to answer American complaints about restrictions on trade with the West Indies. But in essence the Rockinghams differed from Grenville only in as much as they were less provocative exponents of parliamentary rights. Whereas Grenville thought Parliament's right to tax the colonies had to be boldly asserted to avoid being lost, Rockingham believed that it was best not to exercise some rights, or at any rate to exercise them with discretion. In 1766 his administration passed its own Revenue Act. This reduced the charges levied by the 1764 Act on molasses entering the mainland colonies; but it made the charges applicable to all molasses, both foreign and British. Americans seem to have allowed their relief at the cut in the duty to disarm them. They appear not to have noticed that Rockingham was going farther than his predecessor. By treating British and foreign molasses in the same way, Rockingham's administration had abandoned any pretence that it was legislating to regulate the direction of trade within the empire: the 1766 Act was purely a revenue-raising measure, and one of its clauses announced that the money collected would be used to support the army in America.

*

The repeal of the Stamp Act, accompanied by the face-saving Declaratory Act and followed by the little-noticed Revenue Act of 1766, perhaps offered the opportunity for a restoration of imperial harmony. Within a year, however, a new Anglo-American dispute had arisen. Rockingham left office in July 1766, and the new ministry was headed by the Earl of Chatham, who, as William Pitt, had been the leading British politician in the Seven Years' War. From the colonial viewpoint, Chatham's coming to power must have seemed promising. His strong parliamentary speeches in favour of repeal of the Stamp Act marked him out as a friend to America; indeed, he was seen – somewhat unjustly, given Rockingham's role – as the architect of repeal. Yet Chatham's administration did little to live up to his reputation. Chatham himself, overcome with illness, receded into the background. The government, though nominally under the direction of the Duke of Grafton, was in reality leaderless. Those ministers who tackled colonial affairs showed no disposition to be pro-American. Charles Townshend, the chancellor of the exchequer, was already inclined to see the Americans with a jaundiced eye as a result of his experience at the Board of Trade. American resistance to the Stamp Act seems to have convinced him that executive power in the colonies had to be strengthened. In January 1767 he proposed to raise additional revenue in America to relieve the British taxpayer. It soon transpired that he planned to levy duties on glass, lead, paper, paint, and tea imported into the colonies from Britain. The proceeds were to be used to pay 'independent Salaries for the civil officers in North America'.[13] Townshend was sure that the governors, magistrates, and other royal officers had to be freed from the dominating influence of their legislatures – many of which voted their salaries annually – if the colonies were ever to become more amenable to British direction and management.

This was an ambitious objective; but Townshend believed that he could counter American protests. When Benjamin Franklin had been examined by the House of Commons prior to the repeal of the Stamp Act, he had incautiously given the impression that the colonists were opposed only to internal taxation by Parliament, and that they would accept external taxation - taxation, that is, of their overseas trade. Townshend, while denying the constitutional validity of the the distinction, was therefore able to announce that he had devised a system of taxation that would satisfy the Americans. To make the new duties more effective, an American board of customs commissioners was established at Boston.

Meanwhile, the Earl of Shelburne, secretary of state for the southern department, was busy trying to bring the colonial assemblies into line. An attempt by the Massachusetts legislature to indemnify those involved in the Stamp Act riots was disallowed by the British Privy Council. New York was ordered to comply with the terms of the Quartering Act of 1765, which required magistrates to find accommodation and barrack supplies for regular troops whenever requested by the military, and the local legislature to compensate the owners. Shelburne even proposed to make it '*High Treason* to refuse to *obey* or *execute* any Laws or Statutes made by the King with the Advice of Parliament'.[14] Eventually, however, the ministry opted for the milder measure of securing an act to suspend the New York assembly until it fulfilled all its obligations.

Americans responded to the measures of the Grafton ministry in much the same way as they had to those of Grenville's in 1765–6. John Dickinson of Pennsylvania denounced the suspension of the New York asssembly as 'pernicious to American freedom, and justly alarming to all the colonies'.[15] The Quartering Act he declared to be an instrument of parliamentary taxation very similar to the Stamp Act. Concern about the undermining of the colonial legislatures – widely perceived to be the guardians and guarantors of liberty – also lay at the root of American protests against the Townshend duties and the establishment of an American customs commission. Most colonists acknowledged parliament's right to regulate trade; but duties designed to raise a revenue were no more acceptable than the Stamp Act. The use to which Townshend wished to put this revenue made his scheme doubly repugnant. Rendering royal officials independent of local legislatures clearly strengthened the hand of the executive, and made colonial government less accountable. The new customs officials caused dismay for similar reasons: it was imagined that the customs would become a rich field of patronage at the disposal of the executive, and that the executive would be able to use this patronage to determine the conduct and composition of the legislature. A nightmare vision of creeping executive encroachment haunted Americans. Elections would be influenced by the votes of an army of customs officials; assemblymen would be bought by the offer of places in the customs service for themselves or their friends and dependants. Patronage power would in time corrupt a majority in the assemblies, making their members mere creatures of the government. American liberty would be extinguished.

In February 1768, the Massachusetts legislature protested vigorously against the Townshend duties and urged the legislatures of the other colonies to join in co-ordinated action. The following June, the customs commissioners in Boston were attacked by an angry crowd and forced to flee for their lives. Eventually, merchants in Boston, Philadelphia, and New York agreed to a full-scale non-importation campaign; while local artisans determined to shore up the resolve of the merchants by pledging themselves to non-consumption of British imports. And reinforcing this attempt to persuade Parliament to back down was a wish among many Americans to stem the tide of British 'luxury' goods that were thought to be undermining the simplicity, virtue and independence of life in the colonies: 'The Baubles of Britain', T.H. Breen tells us, were believed to be threatening American liberty as much as were parliamentary taxation and a bloated customs service.[16]

The British reaction to American resistance caused further offence. Lord Hillsborough, the hard-line occupant of the newly-created post of secretary of state for the colonies, provoked the colonial assemblies by trying to prevent them from responding to Massachusetts's appeal for common action. To support the customs commissioners, regular troops were despatched to Boston. Their arrival temporarily restored calm; but fresh trouble was not far away. Deep-rooted American fears of military despotism were given substance by the unpleasant realities of the presence of British troops. Bostonians were robbed, assaulted and otherwise ill-treated in a series of well-publicized incidents. Plebian resentment was heightened when off-duty soldiers secured civilian employment by undercutting local wage rates. Tension between the soldiers and town's people culminated in the notorious Boston 'Massacre' of 5 March 1770, when frightened British redcoats fired on a hostile crowd and killed five Bostonians. The 'Massacre' passed into American folklore as an example of the evils of a standing army. Annual commemorations were used by radicals to stimulate further resistance to British authority.

*

Coincidentally, on the very day of the Boston 'Massacre', the new Prime Minister, Lord North, introduced a parliamentary motion for the repeal of all but one of the Townshend duties. The opposition pressed for total repeal, but the cabinet had decided in May 1769 that the tea duty should be retained as a symbol of Parliament's right to tax. American pressure no doubt played its part in inducing this climb-down, though we should note that some of the colonial non-importation agreements designed to coerce the British government became effective only when the decision to repeal most of the duties had already been taken. We might note, too, that the petitioning campaign by British manufacturers and merchants was by no means comparable to the one preceding the repeal of the Stamp Act. Perhaps more important was the general ministerial belief that the duties were fiscally unsound. Not only were they failing to raise a significant revenue, they were also penalizing British exports to the colonies. To tax British manufactures in this way, North is reported to have told the

Commons, 'was to the greatest degree absurd'.[17]

The repeal of all but one of the Townshend duties reduced the intensity of the imperial crisis; but there were still friction points. Ministerial efforts to coerce the assemblies of North and South Carolina caused considerable upset, and rekindled American anxieties about a British conspiracy. The clash with the South Carolina legislature seemed particularly instructive in this regard: the assembly had voted money for the British anti-authoritarian hero John Wilkes, and the British government insisted that the legislature rescind its vote. Renewed discussion of the establishment of an American episcopacy further fanned the embers of colonial discontent. In 1772 the destruction of the naval cutter the *Gaspée* by Rhode Islanders underlined the continuing hostility of Americans to customs enforcement. When the ministry set up a commission of enquiry into the *Gaspée* incident, colonial committees of correspondence were formed to co-ordinate the response to any further action taken by the London government. And when, in the same year, certain key officials in the colonies started to receive salaries funded by the tea duty, Boston radicals produced a comprehensive statement of colonial rights and grievances. Nor should we forget that the British financial crash of 1772 added to American concerns. As British merchants became more nervous about their outstanding loans to colonial planters, the planters grew more and more worried about their debts and more and more uneasy about their loss of commercial autonomy. Fears of economic enslavement reinforced and intensified fears of political enslavement.

When North tried to assist the hard-pressed East India Company – which itself had suffered badly in the crash – he moved the Anglo-American dispute into a new and more dangerous phase. The Tea Act of 1773 abolished British duties on the Company's tea re-exported to America, while obliging Americans to continue paying the duty levied under Townshend's legislation. The aim was to undercut smuggled Dutch tea and allow the Company to dispose of their vast surplus at a healthy profit. But North was not simply proposing to boost the Company's trade. He clearly had an expanded American revenue in mind. When it was suggested that the East India Company would be better served by retaining certain British duties and abolishing the Townshend tea duty, North acknowledged that there were 'political reasons' for keeping the Townshend duty, and suggested darkly that the Americans were 'little deserving favour from hence'.[18]

If North was hoping quietly to secure a greater colonial contribution to the costs of executive government in America, he miscalculated. The Tea Act created vociferous economic victims: smugglers were of course disadvantaged, as were the merchants who were side-lined by the provision allowing the Company to sell its tea directly through nominated agents acting on commission. And American anger was again influenced by fears about the threat posed to the constitutional liberty that the colonists so cherished. They were convinced that the Tea Act was another attempt at parliamentary taxation and the destruction of the independence and vitality of their legislatures. When the Company's tea ships arrived at Boston in December 1773, they were boarded by local people and the tea was thrown into the harbour.

The Boston Tea Party brought forth a set of parliamentary measures usually known as the Coercive or Intolerable Acts. These were directed at Massachusetts, for long regarded by British politicians as the centre of American disaffection. The first punitive response was the Boston Port Act, which closed the port of Boston and moved the customs house to nearby Salem. Only when Boston reimbursed the East India Company would the port be re-opened for overseas trade. An Administration of Justice Act and a Quartering Act were also passed. The first allowed soldiers or other royal officials charged with capital offences to be tried outside the province; while the second increased the powers of the governor and magistrates to quarter troops where they might be needed to put down riots. But the main punishment came in the form of the Massachusetts Government Act, which was intended to reduce the influence of the 'democratic' element in the colony's constitution and strengthen the 'aristocratic' and 'monarchical'. The calling of town meetings was circumscribed, and the governor's authority was significantly increased, not least because the council or upper chamber of the provincial legislature, which had hitherto been elective, was now to be nominated.

Two further measures were widely seen, in Britain as well as America, as related to this coercive programme. General Thomas Gage, commander-in-chief of the army in the colonies, was appointed governor of Massachusetts; and an act was passed relative to the governance of Quebec. The Quebec Act was in fact the outcome of enquiries initiated in 1771, and it could be interpreted as a far-sighted and enlightened attempt to devise a system of government suited to the traditions of a non-English people. But in 1774 it seemed like another attempt to intimidate and overawe the Americans. The territory of the province of Quebec was greatly expanded to incorporate land to the west of some of the old British colonies. The French population was given the right to exercise their Roman Catholic religion, and the Catholic church was permitted to collect tithes. French civil law – which operated without juries – was retained, and, despite earlier promises to the contrary, government was to be conducted through a governor and nominated council, without an elected assembly. Certain taxes were to be raised by the British Parliament.

The parliamentary opposition accepted the Boston Port Act as a necessary response to an attack on private property; but opposition spokesmen were highly critical of the other coercive legislation. The broad consensus among British politicians about America, which had been fraying for some time, now finally split apart. The Massachusetts Government Bill was criticized by the opposition as a violation of charter rights. The Quebec bill met still fiercer resistance. The tiny English-speaking population of Canada, it was claimed, was being denied the rights of Englishmen – notably the right to trial by jury and the right to be taxed only by their representatives – while the French Canadians were being left to languish in their accustomed system of slavery rather than being offered the full blessings of British freedom. The establishment of the Catholic church set alarm bells ringing; it was seen by some parliamentarians as a very sinister development. One opposition MP highlighted what he believed to be the despotic nature of the bill by suggesting that it was inspired by the same arbitrary principles as lay behind Jacobitism: the bill's passage through the Commons on 10 June – the birthday of the Stuart claimant to the throne – was regarded as most apposite.

If the Rockinghamite and Chathamite opposition groups were hostile to the government's legislative programme, the Americans were still more so. North's attempt to cow Massachusetts failed. The Massachusetts Government Act was openly defied. Town meetings continued; the nominated councillors were harassed and most resigned; popular obstruction ensured that the law courts were unable to operate; a provincial convention took over the role of the assembly, which had been suspended by the governor. A public meeting in Suffolk County, in which Boston was situated, called for military preparations and declared that there was no obligation to obey laws that violated the accepted rights of British subjects. Gage's despatches home painted an ever more gloomy picture. By the late autumn of 1774, royal authority had collapsed beyond the environs of Boston, where Gage's army was concentrated. Ensign Jeremy Lister, a young officer newly-arrived from Quebec, wrote on 2 December: 'It is thought by every body here, that we shall be obliged to take the field early next Spring, as the people in general are very enraged against some of the late proceedings of parliament and some determined to defend what they call their Liberties, to the last Extremity'.[19]

In the event of such an armed clash, the British were unlikely to be facing just the people of Massachusetts. Far from being isolated, Massachusetts was backed by the other colonies in an impressive display of American solidarity. Material assistance was despatched to Boston as soon as news of the Boston Port Act reached the colonies; and once the rest of Parliament's American programme became clear, a new Congress, uniting the colonies in one body, was organized. This met in Philadelphia in September 1774, and rapidly endorsed the Suffolk resolves. Shortly after, Congress rejected a conciliation plan put forward by Joseph Galloway that envisaged an American council of all the colonies sharing control of American legislation with Parliament, both bodies having the right of veto. In October Congress declared the Coercive Acts 'impolitic, unjust, and cruel, as well as unconstitutional, and most dangerous and destructive of American rights' and demanded their repeal together with that of other offensive parliamentary legislation.[20] A Continental Association followed, pledging the colonists to suspend all commerce with Britain and to stop the consumption of all British goods pending redress of their grievances. At the same time, the Association – like earlier non-importation and non-consumption agreements – sought to insulate Americans from the corrupting influence of British wealth and sophistication. Indeed, non-importation and non-consumption became the basis for a drive for moral regeneration in which the rejection of luxury and a return to the frugality of simple rustic life would play an important part. Even wealthy Virginia planters felt obliged to temper their aristocratic lifestyle to keep control of what was becoming a powerful grassroots movement.[21]

Given what we know of their earlier hostility to the imposition of parliamentary authority, the colonists' determined resistance to the Coercive Acts is hardly surprising. Indeed, the only real surprise is that North's government imagined that Massachusetts could be separated from the other colonies. Lieutenant-Governor William Bull of South Carolina was convinced that most Americans feared that if Parliament won 'then all the variety of ways and means of raising money in Great Britain will be soon put in practice here and applied to purposes not merely American'.[22] But these fiscal considerations only reinforced and exemplified the main American argument that their own assemblies were the only constitutionally-proper bodies to levy taxation in the colonies. Anything that undermined the independence and integrity of those local assemblies was bound to provoke a fierce reaction. The remodelling of the Massachusetts charter was predictably seen as an especially threatening move, redolent of the campaign to remodel charters in England and the colonies undertaken by the Stuarts before the revolution of 1688–9. A new version of 'Stuart despotism' was again visible to those who had eyes to see it. If Massachusetts could be dealt with in this way, were the other colonies any more secure? The Quebec Act added to the alarm. To

Americans convinced of an unfolding plot to destroy liberty, it was a blueprint for the system of government the British wished to impose all over America. From Philadelphia, Deputy-Governor John Penn reported on 5 September 1774 that people throughout the colonies believed in 'a formed design to enslave America, and though the Act for regulating the government of Canada does not immediately affect the other provinces it is nevertheless held up as an irrefragable argument of that intention'.[23] That Quebec, the former citadel of French despotism, should appear to be providing the pattern was particularly horrifying. It was as if popish tyranny, far from being extirpated with the glorious conquest of New France in 1760, had infected the British. They had succumbed to the temptations of arbitrary power; English liberty was no longer safe in their hands.

*

The imperial crisis was clearly coming to a head; yet even at this stage some Britons believed a peaceful resolution was still possible. Both wings of the parliamentary opposition put forward schemes of conciliation. Chatham in the Lords suggested that a permanent all-American legislature – like the Continental Congress – should exercise the taxing power claimed by Parliament, while Parliament retained its power in other areas. Edmund Burke, speaking for Rockingham's party in the Commons, offered repeal of all legislation offensive to the Americans and a return to the requisition system that had operated in the Seven Years' War. In essence, Burke sketched out a proposed solution based on what would much later be termed 'dominion status', the central feature of which was that Parliament's supremacy should be exercised only when all parts of the empire concurred. Neither of these proposals had much chance of success in the increasingly polarized political atmosphere of early 1775; most British politicians supported North's line, and were determined to make no substantial concesssions to the insubordinate colonists.

North himself, however, also presented a scheme for conciliation. He proposed that the American assemblies raise the taxes to meet the cost of imperial defence and administration, and that Parliament exercise its right to tax only if the Americans failed to produce the requisite funds. Such was the strength of anti-American feeling in Parliament that some of North's supporters thought that he had conceded too much. It seems that only persuasive speeches from the ministerial side reassured uneasy backbenchers. The opposition, meanwhile, claimed that North's offer was a disingenuous attempt to break the united front of the colonies by tempting the more moderately-inclined to come to terms. They were probably correct: it certainly made sense from a government perspective to try to divide and weaken the Americans. North and his colleagues believed that Massachusetts was the prime offender, and that military force would be needed to put down its rebellion. To this end, reinforcements – albeit inadequate to the task – had already been sent to Gage at Boston. If an offer of terms could isolate Massachusetts from the other colonies, so much the better.

In any event, the Americans were no more inclined to accept North's carrot than to bend to his stick. As Franklin commented: 'I cannot conceive that any Colony will undertake to grant a Revenue, to a Government that holds a Sword over their Heads with a Threat to strike the moment they cease to give or do not give so much as it is pleas'd to expect. In such a Situation, where is the Right of giving our own Property freely or the Right to judge of our own Ability to give?'[24] The colonial legislatures, while apparently allowed a role, were in reality undermined as much as if Parliament itself imposed the taxation. After a long delay, Congress finally rejected North's proposals at the end of July 1775, by which time the crisis had escalated into a war.

*

During the winter of 1774–5 the situation in America changed dramatically. In many colonies royal government was in effect replaced by extra-legal provincial congresses and local committees. Bloodshed now seemed inevitable to many Americans. Throughout the colonies preparations were being made for armed resistance to the military forces commanded by North's ministry. 'I think most of the people are mad,' a Scottish visitor remarked, 'in South and North Carolina, Virginia, Maryland, they muster and are every where learning the exercise as if they were going to be attacked.'[25] Rhode Islanders seized the cannons in Fort George in December 1774 and conveyed them to Providence. According to the elected governor, the inhabitants 'meant to make use of them, to defend themselves against any power that shall offer to molest them'.[26]

In Massachusetts the presence of Gage's army made a violent confrontation much more likely. At Salem in February 1775 only the prudence of the commander of a detachment of troops looking for hidden munitions averted a serious clash with local inhabitants. Most officers in the Boston garrison were less inclined to restraint. Captain William Glanville Evelyn of the King's Own Regiment hoped 'we shall shortly receive such orders as will authorize us to scourge the rebellion with rods of iron'.[27] Major John Pitcairn of the Marines wrote that the troops were 'most Willing to give the people here a severe Chastisement', and that his men were 'heartily provoked'.[28] An incident at Roxbury, near Boston, in March illustrates his point. A townsman was reported to have started a brawl with the troops when he lifted up 'the Skirt of his Coat, [and] struck his Hand, as an insult to the Soldiers, on his Backside'.[29]

Even the patient and conciliatory Gage was starting to believe that decisive action on his part was necessary. On 14 April he received orders from London to supress the rebellion. He decided to strike against the local militia magazine at Concord. Late on the night of 18 April, a force of 800 troops under the command of Lieutenant-Colonel Francis Smith of the Tenth Foot left Boston for Concord. Early the next morning, Smith's troops were confronted by militiamen on Lexington Green. The regulars ordered the militia to lay down their arms and disperse. Someone – no-one knows who – let off a musket. The regulars, believing themselves to be under attack, opened fire. Eight militiamen were killed. Smith then pushed on to Concord, where he destroyed what munitions he could find. Parts of the town were set alight. The local militia resisted and the regulars began to retreat towards Boston. Very soon they were facing a devastating assault from hidden militiamen along the route. According to Ensign Henry de Berniere of the Tenth Foot: 'we at first kept our order and returned their fire ... but when we arrived within a mile of Lexington, our ammunition began to fail ... we began to run rather than retreat in order'.[30] Smith's troops would almost certainly have been compelled to surrender had it not been for the timely arrival of a relief force under Brigadier-General Earl Percy. Percy's brigade, including artillery, held the enraged militiamen at bay, allowing Smith's command to rest and regroup. The British then resumed the retreat to Boston. But their sufferings were not over. 'We had not marchd 200 yards,' an officer in the relief column recalled, 'till we received a hot fire from all sides ... every wall lined, and every house filled with wretches, who never dared to shew their faces.'[31] At Menotomy more militia arrived to harass the regulars, and from there to where the British left the Cambridge road to take the back route to Charlestown the fighting was particularly fierce. Percy reported that the troops had 'retired for 15 Miles under an incessant fire which like a moving Circle surrounded & followed us wherever we went'.[32] By the time they reached the safety of Charlestown, across the harbour from Boston, the British had suffered 273 casualties. After months of mounting tension, the war had begun.

Chapter I Notes

1. The backgrounds to the Anglo-French, Anglo-Spanish, and Anglo-Dutch dimensions are discussed in Chapers 3 and 9.
2. Quoted in Beckles Willson, *The Life and letters of James Wolfe* (London, 1909), p.392.
3. Jared Ingersoll to Thomas Fitch, 6 March 1765, 'Fitch Papers', Connecticut Historical Society, *Proceedings,* xviii (1920), 334.
4. *Prologue to Revolution: Sources and Documents on the Stamp Act Crisis, 1764–1766,* ed. Edmund S. Morgan (New York, 1959), p.28.
5. *English Historical Documents: American Colonial Documents to 1776*, ed. Merrill Jensen (London, 1969), p.678.
6. See Paul Langford, 'Property and "Virtual Representation" in Eighteenth-Century England', *Historical Journal,* xxxi (1988), 83–115; Joanna Innes, 'Parliament and the Shaping of Eighteenth-Century Social Policy', Royal Historical Society, *Transactions,* 5th series, xl (1990), 63–92.
7. *The Regulations lately made Concerning the Colonies and the Taxes imposed on them Considered* (London, 1765), pp.108–9.
8. *The Rights of Colonies Examined* (Providence, R.I., 1765), p.16.
9. *The Diary of Colonel Landon Carter*, ed. Jack P. Greene (2 vols., Charlottesville, Va., 1965), ii. 821.
10. P.D.G. Thomas, *The English Satirical Print 1600–1832: The American Revolution* (Cambridge, 1986), pp.48–9.
11. Newcastle Papers, Add. MS 32,699, fo. 257.
12. Rockingham to the Duke of Newcastle, 2 Jan. 1766, Newcastle Papers, Add. MS 32,973, fo. 11.
13. Townshend to Grafton, [25 May 1767,] Grafton Papers, Ac 423/445, West Suffolk RO.
14. Shelburne to Chatham, [8 March 1767?,] Chatham Papers, PRO 30/8/56, pt. i, fo. 89.
15. *Letters from a Farmer in Pennsylvania* (Philadelphia, 1768), Letter I.
16. See Breen's '"The Baubles of Britain": The American and Consumer Revolutions in the Eighteenth Century', *Past & Present*, no. 119 (1988), 73–104; and 'An Empire of Goods: The Anglicization of Colonial America', *Journal of British Studies*, xxv (1986), 467–99.
17. *Proceedings of the British Parliament regarding North America, 1754–1783*, ed. R.C. Simmons and P.D.G. Thomas (6 vols. to date, London, 1982–), iii. 228.
18. *Proceedings of the British Parliament*, iii. 492.
19. Lister to Col. William Fawcett, 2 Dec. 1774, Shibden Hall Muniments, Calderdale District Archives.
20. *American Colonial Documents*, pp.806–8.
21. See Rys Isaac, 'Evangelical Revolt: The Nature of the Baptists' Challenge to the Traditional Order in Virginia, 1765 to 1776', *William & Mary Quarterly*, 3rd series, xxxi (1974), 345–68.
22. Bull to Dartmouth, 31 July 1774, *DAR*, viii. 153.
23. *DAR*, viii. 186.

24. *Franklin Papers*, xxi. 510.
25. *Travels in the Colonies in 1773–1775*, ed. Ted Ruddock (Athens, Ga., 1993), p.75.
26. *Naval Documents of the American Revolution*, ed. W.B Clark and W.J. Morgan (7 vols., Washington, D.C., 1964–76), i. 15.
27. *Memoir and Letters of Captain W. Glanville Evelyn*, ed. G.D. Scull (Oxford, 1897), p.51.
28. Pitcairn to Sir Roderick Mackenzie, 16 Feb. 1775, Mackenzie Papers, Add. MS 39,190, fo. 209.
29. Examination, 21 March 1775, Gage Papers, William L. Clements Library.
30. [Henry de Berniere,] 'Gage's Instructions and Narrative', Massachusetts Historical Society, *Proceedings*, 2nd series, iv (1816), 217.
31. British Army Journal, 1775–7, p. 12, Henry E. Huntington Library (photocopy in New York Public Library).
32. Percy to Gen. Edward Harvey, 20 April 1775, Percy Papers, Vol. L, Pt. A, fo. 53, Alnwick Castle, Northumberland.

rate from the Seven Years' War. But a further and more serious qualification to the enthusiasm of Goethe and Clausewitz is in order. The developments that came to fruition in Continental Europe in the 1790s had already appeared, in almost the same state of ripeness, across the Atlantic. It was in the War of American Independence that the defining features of the French Revolutionary War were most fully anticipated. If we have to pin-point a time when the old gave way to the new, it was 1775 not 1792.

*

Behind this assessment – as behind so many historical judgements – lies a mass of contradictory evidence. It would be an easy matter to assemble material to support the opposite view. France entered the war against Britain in 1778, Spain in 1779, the Dutch in 1780. It would be stretching credulity beyond its limits to claim that the armies of these European powers were in any sense revolutionary. Their officer corps were predominantly aristocratic; their common soldiers were taken from the lower strata of society, including criminals and beggars. Their armies relied heavily on foreign mercenaries: the Dutch had their German troops and a Scots brigade; the Spanish their regiments of Flemish, Walloon, Italian, Irish and Swiss troops; while in the French army nearly a fifth of the line regiments were foreign – Swiss, Irish, Italian or German. True, the Spanish and French introduced reforms in response to their defeat in the Seven Years' War. But these reforms did little to alter the *ancien régime* character of their military forces. Indeed, in some respects they reinforced it. In Spain the abandonment in 1776 of the *sorteo*, a system of conscription by ballot, increased dependence on the *leva*, a levy of prisoners and vagrants. In France, although a few daring officers – of whom Jacques de Guibert was the most well-known – recommended the creation of a truly national army by resurrecting the classical ideal of the citizen-soldier, the most influential reformers placed their faith in a reassertion of the values of the existing order. France would be restored to her former glory, they argued, only when she recovered her martial qualities. Since these martial qualities were essentially aristocratic qualities – leadership, courage, honour, chivalry – it followed that what was needed was a purge of *bourgeois* infiltrators. In 1781 these views found expression in the Ségur law, which required officer vacancies to be filled only by those who could prove that all four of their grandparents were of noble origin.[4]

The British army could also be seen as a typical military institution of the *ancien régime*. Its officer corps was well stocked with aristocrats and members of the landed gentry like Hugh, Earl Percy, heir to the dukedom of Northumberland; Alexander Lindsay, Earl of Balcarres; John Sinclair, Earl of Caithness; and John Dyke Acland, the son of Somerset baronet Sir Thomas Dyke Acland. Common British soldiers, in marked contrast, were often represented as coming from the bottom of the pile: Major-General Robert Cunninghame described a set of Irish recruits as 'the very Scum of the Earth'.[5] In the ranks were paupers compelled to serve under the provisions of the Press Acts of 1778 and 1779; the likes of Daniel Vescoy, a 'Rogue & Vag[abond]' released by Warwickshire magistrates on condition that he joined the Marines;[6] and thieves such as William Smith, detained for stealing a bag of tools and 'Discharged on his ... inlisting into his Majesty's Service, as a Soldier'.[7] The British forces were also noticeably cosmopolitan. Some of the officers and men in the British regiments were from abroad. The Royal Americans – despite their name – were mainly German and their officers included members of the Prevost family, Swiss soldiers of fortune who had sold their services to many masters: Augustine Prevost had started his career as an officer in a Swiss regiment of the Dutch army. At the start of the American war, Georg von Scheiter, a Hanoverian colonel, offered, as a business proposition, to raise 2,000 recruits in Germany. The men he finally assembled were used to top up under-strength British regiments. More importantly, the regular British troops were augmented by whole units hired from German princes by the British government. Hesse Cassel alone provided nearly 19,000 troops during the course of the war. In 1778 German auxiliaries accounted for 33% of the Royal army in America; by 1781 the proportion had risen to 37%.

Even the American Continental army bore more than a passing resemblance to European armies of the age. Its officers included substantial landowners – George Washington, the commander-in-chief, was a notable example, as were the New York patroons, Philip Schuyler and Philip van Cortlandt. The rank and file was drawn mainly from the poorest sections of colonial society, even those who languished at its very bottom. Washington described the New England soldiers besieging Boston in 1775 as 'an exceedingly dirty and nasty people'.[8] The number of poor and needy Americans in the ranks probably increased as the war dragged on: shortages of volunteers led to conscription, and more comfortable citizens, when drafted into the army from the militia, often preferred to pay substitutes to serve in their place. These substitutes were usually the most disadvantaged members of local communities. In recognition of the difficulty of persuading many prosperous Americans to serve, in March 1779 Alexander Hamilton recommended the large-scale enlistment of blacks on the grounds that 'their want of cultivation' and 'that habit of subordination which they acquire from a life of servitude, will make them sooner become soldiers than our white inhabitants'.[9] There was some reluctance to adopt Hamilton's proposal; in the south, where slaves formed a majority in some areas, the prospect of potentially-mutinous armed blacks caused particular alarm. Nonetheless, in the northern contingents of the army, at least, blacks played a significant role in reducing the severity of the manpower problem. In 1781 a French officer inspecting the Continentals at White Plains, New York, noted that 'A quarter of them were negroes.'[10] The Continental army's ranks also included deserters from the British and men known to be disaffected to the revolutionary cause who were forced to spend time with the colours as punishment. This perhaps explains the fierce discipline imposed in the American army; a fierce discipline that seems strangely at odds with the image of a revolutionary people contending for their liberty. The British army may have been harsher – regularly administering 1,000 lashes, and even 1,500, compared with the official American limit of 100 – but the full range of sanctions employed by contemporary European armies was used by Washington's officers. Journals and order-books refer to the infliction of humiliation ('his punishment was to ride a wooden Horse ten Minutes')[11] pain ('Wm Dearlove ... Sentenced to receive 100 Lashes well laid on upon his bare back')[12] and even death ('This afternoon was executed ... David Hall of the Pensilvania line – for robing an Inhabitant').[13]

The conduct of the war, as well as the nature of the armies fighting it, could be portrayed as characteristic of the pre-French Revolution period. The Anglo-French conflict seems to have been fought not only for traditional ends – to secure territorial, colonial and commercial advantage – but also in a traditional (and decidedly gentlemanly) manner. Major John Bowater of the Marines, held prisoner in Brittany at the end of 1778, was given freedom to hunt, shoot and fish within six miles of his place of detention, and he was able every day to buy a three-course meal 'with Claret & white wine'. Small wonder he considered himself 'unfortunate' but 'not unhappy'.[14] While the French were trying to retake St Lucia in 1779, Colin Lindsay, an officer in the Forty-sixth Foot, noticed how their sentries 'when they saw our soldiers passing near, would point to their arms, shake their heads and laugh, but never fired'.[15] Major William Saxton, a British army officer convalescing in France before the signing of the Franco-American alliance, saw no reason to cut short his enjoyable soujourn once Britain and France became enemies.

In America, too, old regime norms seem regularly to surface. For long periods the British commanders failed to bring the Americans to action; and when they did manage to engage the revolutionary forces in battle, they seemed reluctant to follow up their successes. Most senior British officers, moreover, appear to have wished to avoid a policy of large-scale destruction; indeed, at times they went out of their way to limit the disruption caused to civilians. Earl Percy refused to order the seizure of most of the hay when the British occupied Rhode Island, on the ground that 'it is absolutely necessary to leave at least 800 Ton for the Stock of the Inhabitants, without which, they must be reduced to the greatest State of Misery'.[16] The approach of the senior British officers in America seems, in short, to exemplify the timidity, caution and – more positively – the civilized moderation associated with pre-French Revolutionary warfare.

On the American side there were also those who wanted to fight according to the established rules of European etiquette. Washington envisaged the Continental army as a regular force that could be made to match the British army. He consciously emulated British methods, training his men to adopt battlefield formations that would not have looked out of place on a European parade ground. A conservative Virginia planter who had dreamed of a commission in the British army, Washington regarded the American militia with deep suspicion – especially the New Englanders, with their habit of electing officers. He was implacably opposed to a full-scale guerrilla war, the radical alternative urged by his colleague (and rival) Charles Lee.

*

Yet, for all this, one should be wary about falling for the attractions of the *ancien régime* model. America, as we shall see in a later chapter, was greatly affected by the war; her economy, society and politics were all changed by the experience of a protracted struggle. And the American military effort cannot really be compared with the Spanish and the French. Washington may have distrusted and criticized the militia, but militiamen played a vital role in the struggle. The militia acted on occasion as a political police, rounding up and intimidating local people who remained loyal to the British. But it contributed as a fighting force as well. Militia units served alongside Continentals – not always, it has to be said, very successfully – and outside the set-piece engagements they made a significant contribution, discomforting their enemies by fighting in a style that can only be likened to guerrilla warfare. The British regularly faced popular resistance on a scale and of an intensity no more than glimpsed in previous eighteenth-century conflicts. On the retreat from Concord, as we have seen, the redcoats were almost overwhelmed by a hostile populace. According to one report, 'even Weamin had firelocks: one was seen to fire a Blunder bus between her Father, and Husband, from their Window'.[17] Two years later, in April 1777, a British officer in New Jersey observed that the Americans 'take every opportunity, of firing, upon our people, from behind fences, trees, &c'.[18] That summer a British force advancing into New York from Canada also met with determined opposition – much of which was provided by militiamen. A bewildered British officer wrote that 'the Country engenders Rebels, and as we pursue them in Front they arise afresh in the rear'.[19] In 1780 the story was no different. The British adjutant-general in America complained of 'the perfidious Thousands of the Jerseys & of New England who singly take the field against us from every tree & house & who tho driven from front close in with redoubled inveteracy on our Flank & rear'.[20]

The militia was revolutionary in another sense. It contained Americans from all backgrounds: in fact nearly every free male of military age was eligible for service. Not everyone was required to turn out, of course; and richer militiamen, rather than actually fight, could hire substitutes. Indeed, it seems that some poorer colonists made their living as perpetual substitutes – serving in place of one man one year and another the next. But not every American who could afford to buy his way out of active service chose to do so; and at times even those who wanted to hire a replacement found that the stock had been exhausted: 'no Substitutes can be procured', Virginian planter Robert Beverley moaned to his son, 'This being the Case, Conceive to yourself your Father in the dignified Character of a Common Soldier'.[21] And if the militia ranks were manned by a genuine social mix, its officers – especially in New England – were by no means always drawn from the elite. The novelty of this to foreign eyes is worth noting. On the march in New England in May 1781, one of the aides to the French commander-in-chief observed: 'Our innkeeper was a captain, the several military grades being granted here to every rank of people. There are shoemakers who are colonels', he continued, adding with evident amusement, 'it often happens that the Americans ask the French officers what their trade is in France.'[22]

Nor was it just the American militia that was revolutionary. On closer inspection any resemblance of the Continental army to typical armies of the *ancien régime* starts to fade. Upper-class officers like Washington, Schuyler and van Cortlandt rubbed shoulders with colleagues from more humble backgrounds. Before the war, General Nathanael Greene had been a Rhode Island ironmaster; General Daniel Morgan had spent time as a wagon driver. During the conflict hundreds of Americans shared the experience of Ebenezer Wild of Massachusetts, Thomas Anderson of Connecticut, John Ashton of Pennsylvania, Burgess Ball of Virginia, and Robert Barnett of New Hampshire, all of whom became officers after having served in the ranks. To a much greater extent than any other army of the time, the Continental army embodied the principle – usually associated with the French Revolution – of careers open to talent. The ordinary American soldiers can likewise be distinguished from their European counterparts. In the New Jersey line, more than 40% of the rank and file owned land – a much higher proportion, surely, than in any European army.

The American armed forces were not only more socially representative and more meritocratic than other armies of the age; they were also larger, in relation to population. By the end of the war, perhaps 200,000 American males had engaged in some kind of military service – as a Continental, in a state regiment, in the militia, or on board a warship. If we use the total population – white and black – as our starting point, we can calculate that this represented about one in every three to four men of the appropriate age. When we take account of the American loyalists who could not be called upon – perhaps a fifth of the population – and the reluctance of most states to allow large-scale slave enlistment, the proportion goes up to nearly one in two. In some communities the military participation rate was higher still. Peterborough, New Hampshire, for instance, had a population of 549 in 1775. About 170 of its men performed some kind of military service during the struggle for Independence. New Ipswich, in the same state, was just as impressive, providing 360 men out of a total population of 1,033; and Malden, Massachusetts, with 1,030 souls in 1776, fell only marginally short of mobilizing its full manpower potential when it sent 231 men into the Continental army or militia. These places might, of course, have been exceptional. The staunchly Protestant New England colonies had contributed many of their men during the Seven Years' War for the great crusade against Catholic Canada: they could almost be said to have had a tradition of significant mobilization of manpower. But between 1775 and 1783 a high proportion of eligible males in other states also served. More than two-thirds of the men aged 16 to 50 in Chatham, New Jersey, spent time as a militiaman, in a state regiment, or as a Continental soldier. In Guilford township in Cumberland County, Pennsylvania, one man in every two of the appropriate age was actually serving with the militia in 1779; and this does not count others who might have been away with the Continental army. A recent calculation suggests that in Virginia 'Around half of white men over sixteen saw active duty either in the army or in a militia unit activated for state defence or national service.'[23] In many places the numbers involved were so great that even war-related industries were disrupted: in New Jersey a manufacturer of gunpowder complained that 'all the Men I Depended on to work the Mill are called out from my Service in the Militia'.[24] In short, we know enough about the general pattern of mobilization in the War of Independence to be able to say that in numerical terms the Americans became a nation in arms more than 15 years before the French.

True, numbers are just one ingredient in the recipe. A nation in arms needs to identify with the nation's cause; soldiers need to share a common purpose, to be motivated – at least in part – by the same ideals. We can say that the British saw their opponents as buoyed up by a new kind of fervour: in June 1775, just after the sanguinary battle of Bunker Hill, General Thomas Gage, the British commander-in-chief, declared that the Americans 'shew a spirit and Conduct against us they never shewed against the French'.[25] Even so, the level of ideological awareness in the American ranks is not easy to measure. Ordinary soldiers, in particular, rarely left a record of their political thoughts. Their diaries, when they survive, tend to to be dominated by accounts of weather conditions, distances marched, and food consumed. Religious comments are more frequent than political ones: Americans saw themselves as fighting for God; and occasionally we can see evidence of a millenarian motivation.

But we should not assume from the use of religious language that for many Americans the war was not a political struggle. For participants on both sides, spiritual and political agendas were often inseparable. On one level this meant that revolutionaries believed that Christ's second coming might be ushered in when the Americans secured their freedom. On another level it meant that loyalist clergymen like Henry Caner could write of the 'same levelling principles' that caused Americans to 'withdraw from the wholesome establishment of the Church' and to throw off 'the restraints of civil Government',[26] while revolutionaries could argue that High Anglicanism of Caner's kind, with its emphasis on religious hierarchies and the doctrine of passive obedience to the powers that be, was ultimately irreconcilable with political liberty.[27]

Indeed, if one looks carefully enough, and pieces together scraps of evidence, a picture emerges of what appears to be a deep – if by no means universal – commitment to the Revolution on the part of both officers and men in the American forces. Baron von Steuben, Washington's drillmaster, explained to a European friend: 'You say to your soldier, "Do this," and he doeth it, but I am obliged to say, "This is the reason why you ought to do that," and he does it.'[28] American soldiers, as von Steuben appreciated, were unwilling to be turned into unthinking automata. In 1777, as Washington's troops marched through Philadelphia, John Adams noticed 'They dont all of them cock their Hats – and such as do, dont all wear them the same Way.'[29] Such defiance – such self-conscious individuality – indicates that many common Americans were anxious not to lose the character of citizen in their search for that of soldier. These anecdotes are of course merely suggestive; they reveal, one might argue, a sturdy anti-authoritarianism rather than political commitment. But we have already noted that hostility to 'passive obedience' was central to the thinking of many revolutionaries: the American battle flags bearing a coiled snake and the motto 'Do not tread on me' provide a graphic illustration of its importance. Here we might observe the way in which American soldiers of all ranks routinely called pro-British Americans 'Tories'. To those reared in a Whig and Protestant tradition, suspicious of executive power and ever fearful of the threat to liberty (political and religious) their adversaries naturally appeared like the Tories of the late-seventeenth and early-eighteenth-centuries – supporters of High Anglicanism and believers in the prerogatives of the crown and obedience to the powers ordained by God.

Washington, for all his aristocratic disdain, had little doubt that his troops were actuated by the same political hopes and fears as himself. When his troops behaved badly he referred to 'the Common cause' and 'the Cause in which we are engaged'. There was, in his view, a sense of shared values on which he could draw. When he was particularly angry, and trying to stop plundering, he spelled out the meaning of the cause: 'Why did we assemble in Arms', he asked rhetorically, 'was it not to defend the property of our Countrymen?'[30] For the most part, however, he needed only to refer to the cause: probably because it seemed pointless to elaborate on what everyone took for granted. And, in support of this point about a commonly accepted language and set of assumptions, we have the simple testimony of at least a few humble Americans in arms. The first entry in the journal of Bayze Wells of Connecticut, 10 May 1775, reads: 'I inlisted my Self A Sergent ... in the fourth Regement Ras^{d} by the United Colonys for the Defence of the same Against the Tireny of Great Britton.'[31] Such enthusiasm may well have been at its strongest in the first wave of resistance to the British: the degree of commitment was certainly to vary as the struggle dragged on. War-weariness was particularly conspicuous in Virginia in the last two years of the fighting. Even so, if language is any guide, revolutionary fervour was present in some measure everywhere in the 13 colony-states throughout the conflict. For Samuel Tirrell, a militia officer from Virginia, the British were 'merciless Tyrants';[32] for Colonel Samuel Blachley Webb of Connecticut, the enemy were 'these wretches the Servants and Slaves of George the Third';[33] while for John Glover, an officer from Massachusetts, the choice facing his countrymen was crystal clear: if they lost, 'which God forbid, we must be slaves, which is worse than Death'.[34]

We have already seen what the Americans meant by slavery and tyranny. We have seen, too, that their arguments – like those of Britons who saw the colonists as Cromwellian fanatics and levellers – have a backward-looking, seventeenth-century religio-political flavour. When Americans talked and wrote of a British plot to deprive them of their liberty, they readily compared it with the authoritarian intentions imputed to the Catholic or crypto-Catholic Stuart monarchs. Yet however historically-rooted the arguments, they acquired an added dimension – a democratic dimension – in the American context.

The franchise in the colonies was considerably wider than in Britain because land, the basis for voting rights in both Britain and America, was held by a larger proportion of the people. The defence of representative institutions in the colonies against the claims of parliament therefore involved a great many Americans who had a personal stake in the continued vitality of those institutions. And as the dispute with the British parliament developed and intensified, more and more humble Americans were drawn in and politicized. On the eve of the war Henry Laurens, a wealthy South Carolinian, wrote incredulously of the way in which local committees, manned by ordinary Americans, were now making key decisions. One such committee, acting on authority given to it by South Carolina's provincial congress (of which Laurens was president), was even determining whether creditors could sue for the recovery of debts: 'figure to your Self,' Laurens wrote to a friend, 'the Sight of the most respectable Characters in Carolina Standing up uncovered & Soliciting certain persons who have no visible property for permission to act according to Law'.[35] Colonists who had been content merely to vote, and had left local government to their 'betters', began to flex their muscles and to challenge the right of their social superiors to dominate the political process.

It must be said that nearly everywhere this new assertiveness was in some measure contained by members of the traditional elite. The state constitutions that were formulated from 1776 tended to lean heavily towards the Whig model of checks and balances; only a few gave full expression to the radical demand for democratic control, and the most celebrated of these – Pennsylvania's – was revised in a more conservative direction in 1790. After the conflict was over Gouverneur Morris, an upper-class New Yorker who had been very apprehensive about plebian participation in politics, was able to breathe easily again: 'I think the superior Advantages of our Constitution will now appear in the Repressing of those turbulent Spirits who wish for Confusion'.[36] But even if the democratic surge was resisted and in many places neutralized, it was an important feature of the Revolution. To adapt a famous formulation, what started as a struggle for home rule became, in part at least, a struggle to decide who should rule at home.

It seems, then, that a participatory political culture, deepened and extended by the Anglo-American constitutional crisis, and still more by the war itself, helps to explain why so many Americans were willing to fight. They were defending what they held dear – their liberties, and, more precisely, their political involvement through their representative institutions. But many ordinary Americans who took up arms were not satisfied merely with protecting the colonial dispensation against British encroachments; they sought to seize the opportunity to create a new order in which their own influence could be more effectively exerted. Popular military participation and the inspirational ideology of popular political participation, the hallmarks of Goethe's European 'new era' ushered in at Valmy, had already joined in potent union in revolutionary America.

*

What about the enemies of the American revolutionaries? It may be possible to construct a good case for novelty on the American side, but surely not on the British? Improbable as it might seem, such a case can be made. Even when we focus on the British – apparently the conservative defenders of the *status quo* – the struggle seems to depart in a number of important respects from the 'limited' war paradigm. Despite appearances to the contrary, the British commanders in America wanted to knock out the Continental army: indeed, they saw this as the key to a successful conclusion of the conflict. 'I am still of opinion that peace will not be restored in America until the rebel army is defeated', wrote General William Howe in July 1776.[37] Howe was perhaps held back at various times in the ensuing campaign by a desire to avoid ruining the chances of a negotiated settlement: there is some evidence to suggest that, under the influence of his brother, Admiral Lord Howe, he failed to seize every opportunity to crush his enemy. But General Howe seems to have believed that peace could be obtained only once there had been a clear demonstration of British military might; like his brother, he thought that negotiations were most likely to follow a decisive British triumph on the battlefield. Even when he had given up hope of a compromise settlement, he spent much of his time seeking a set-piece action with Washington. So did Howe's successor, Sir Henry Clinton. They were inhibited in their quest not by any eighteenth-century delicacy about avoiding pitched battles, but by the peculiar conditions of the American war. Three thousand miles of ocean separated them from their chief source of supplies and reinforcements, and naturally inclined them to minimize risks in the search for victory. The loss of the main royal army in America would have been disastrous for the British cause, not least because it would have convinced undecided and even friendly colonists that the British were unlikely to win.

Likewise the desire of the British commanders to limit the sufferings of civilians owed much to a realization that they were fighting a distinctly unconventional war. Colonial goodwill was the prize at stake: 'to gain the hearts & subdue the minds of America' was how Clinton described the task.[38] We should also bear in mind that, even if generals like Clinton sought to restrict destruction for this reason rather than any attachment to a generalized concept of 'limited' war, less senior officers often had no time for restraint of any kind. 'Lenity will not do', Captain Harry Rooke of the Fifty-second Foot announced briskly.[39] There were many officers of junior and middle rank like Rooke who urged the adoption of severe methods. More to the point, some of them had the chance to undermine the efforts of the commanders-in-chief. Advocates of 'fire and sword' could find themselves in charge of detachments and able to translate their rhetoric into terrifying reality. Whatever the wishes of senior officers like Clinton, much of the war at grassroots level was far from restrained.

We might also note that, despite Major Bowater's apparent lack of concern at his detention in Brittany and the old-world courtesies observed on St Lucia, the outbreak of hostilities between Britain and France inspired a wave of patriotism comparable to that seen in the 1790s. Deep-rooted antipathy to the French was given extra bite by resentment at what was seen as their interference in a domestic quarrel – a civil war within the British empire. As the vicar of Hanbury in Staffordshire noted on 1 January 1779: 'In consequence of the Success of rebellion the ever-perfidious *French* have, this last year, (1778) joined the Americans'.[40] And so far as the insouciant Major Saxton is concerned, his conduct was by no means typical: in fact the war office in London considered him a scandalous reprobate.

This was partly, perhaps, because the British army, notwithstanding its traditional features, was not the same as the Spanish and French armies. First of all, its cosmopolitanism needs to be put in proper perspective. German soldiers were employed in large numbers in North America, but they were much less conspicuous elsewhere. Although a few Hanoverian regiments formed part of the garrisons at Gibraltar and Minorca, and two units even went out to India, no foreign auxiliaries were used in the Caribbean or in the British Isles. Taking account of all theatres of operation, the proportion of foreign troops in British pay was smaller than in any previous eighteenth-century conflict.

Just as important, by no means every British officer was from a landed background. James Robertson, who rose from lieutenant-colonel to lieutenant-general in the American war, is a case in point. He managed to marry his daughter to the son of a Scottish baronet, and as a result of this match he found himself after the war exerting an influence in the narrowly exclusive world of Scottish electoral politics; but Robertson himself sprung from decidedly humble origins in Newbiggin, Fife. He is said to have joined the army as a private. Lieutenant George Dunlop was the son of a Scots merchant; Captain Frederick Mackenzie's father was a Dublin tradesman; Lieutenant Gilbert Waugh's was reputed to be an Edinburgh baker. James Green, a lieutenant in the Twenty-sixth Foot, started his wartime career as a sergeant.

Nor was the rank and file made up of paupers and criminals. The haul of wandering beggars conscripted under the provisions of the wartime Press Acts was unspectacular: from Norfolk Captain John Money of the Ninth Foot reported in March 1779 'that only seven Men have yet been inlisted on the Impress Act & only two from 150 parishes'.[41] By halfway through the life of the 1779 Act no more than 1,463 men had been pressed in England and Wales, and a paltry 67 in Scotland between March and July of that year. Even these figures exaggerate the number taken into the army, for some of these unwilling recruits injured themselves to avoid active service: a return of impressed men destined for the Ninety-ninth Foot in 1780 includes several entries that read: 'lost his right Thumb'.[42] The number of criminal entrants appears also to have been small, not least, as an under-secretary of state noted, because of the frequent 'Refusal of the Officers to admit them'.[43] Most offenders were sent to the disease-ridden tropical climes to spare troops of better character; serious malefactors were packed off to the African Corps and almost certain death in Senegambia. The bulk of the British rank and file entered by a different route, and seem less like members of the permanent underclass brought to mind in historians' accounts of *ancien régime* armies. Many, admittedly, had been unskilled labourers in civilian life, but out-of-work artisans and even merchants and schoolteachers down on their luck could be found with the colours. Although skilled men had joined the army before the American conflict, wartime changes in the methods and conditions of recruitment probably attracted more than ever. In 1775 the time of enlistment was reduced from an indefinite period to three years or the duration of the emergency; while augmentation of bounty money paid to new entrants must have increased temptation for those suffering from acute but perhaps unaccustomed distress. Here and there men 'of condition & property, Sons of freeholders &c' are identified in contemporary recruiting correspondence, and on one occasion even the son 'of a reputable Clergymen'.[44] Most British soldiers were poor and needy when they joined the army, but the rank and file represented a broader cross-section of society than the *ancien régime* stereotype would suggest.

Military recruitment in Britain was reaching out beyond the traditional pool of upper- and lower-class manpower – and a smaller proportion of foreign troops was being employed – because the number of Britons under arms was larger than ever before. True, in the early stages of the war, the British were slow to devote adequate resources to the struggle. But every effort was made once the French entered the fray. Indeed, although the contest with the Bourbons was in many senses 'traditional', its impact on British mobilization can be more appropriately described as 'revolutionary'; it pointed the way to the 1790s rather than simply repeated the experience of earlier eighteenth-century conflicts. By the close of the war more than a quarter of a million British and Irish subjects of King George were serving in the army, navy or militia (110,000 in the army, 107,000 in the navy, and about 40,000 in the English and Welsh militia and Scottish fencibles). It seems likely that another 60,000 or so Irish Protestants were active in unofficial volunteer formations by the middle of 1780.

It would be more helpful, of course, to know how many men served – for however short a period – during the war, rather than how many were serving at any given moment. We need to take account of the turnover of men entering and leaving the armed forces. Between September 1775 and September 1780 more than 73,000 recruits were raised for the army. For the whole war the figure is probably in excess of 100,000. Add this to the actual (as opposed to establishment) strength of the army in 1775 – say 36,000 officers and men – and we have about 136,000 soldiers. For the navy the numbers were larger. Starting with around 16,000 sailors in 1775, perhaps some 171,000 were added during the course of the war (70,000 or so were lost through death or desertion, and others were disabled from active service). Our estimated 136,000 soldiers and 187,000 sailors gives us 323,000 men. Then there is the militia to consider. Not many militiamen deserted or were killed – they saw no action against the enemy – but there were deaths through disease: in November 1779 the two main militia camps in southern England were said to be 'very sickly';[45] while in the summer of 1782 'a violent fever ... carryed off' several members of the Northamptonshire regiment.[46] Moreover, militiamen were required to serve only three years once their units had been embodied, which meant that in theory a whole new complement would have been raised in 1781 and 1782. In practice, it seems that many of the militiamen agreed to serve a second term on the payment of a bounty. Lack of hard evidence makes it especially difficult to reach a figure for the number of militiamen who served during the war; but 60,000 is probably not too far wide of the mark. If we say that for the Irish volunteers 80,000 is probably about right, we have a total, for all branches, of some 463,000 men.

We have still not finished, however. This figure takes no account of all sorts of miscellaneous extras such as the Devonian merchant seamen temporarily 'Entered on board [a] man of war' when the French fleet appeared off New York in 1778,[47] or the English merchant who donned a uniform of 'Scarlet faced with Blue' to help defend the same place.[48] More importantly, we have not considered the unknown number of volunteers raised in Britain rather than Ireland. As in the French Revolutionary and Napoleonic Wars, all manner of volunteer units were formed to repel a feared invasion. Towards the end of the conflict, Mary Hardy, a Norfolk farmer's wife, recorded in her diary that her husband and children had gone to a neighbouring village 'after Church to see the Volunteers exercise'.[49] We know from other sources that in 1778 a volunteer corps was formed in Bath. And, revealingly, a militia officer stationed at Plymouth in 1779 declared: 'There is scarce a Town or Village in this County but what have Raised Independent Companys or Company'.[50] By the end of that year there were said to be about 150 companies of volunteers in England, of at least fifty men each. In 1780 a local newspaper mentioned a volunteer corps in Ipswich, and by 1782 – perhaps earlier – Birmingham had its own Independent Volunteers. In Scotland 'voluntary Associations like the Irish'[51] were formed in 1779–80. They were discouraged by local army commanders, but further units seem to have emerged at the end of the war when the idea of a Scottish militia was seriously canvassed: there are references in contemporary records to the Edinburgh Defensive Band and the Caledonian Band, the last of which seems to have included 'the Earl of Buchan and several other Gentlemen'.[52]

If we say, very roughly, that 30,000 covers all these miscellaneous extras (this might well be a serious underestimate), that gives us only a little short of half a million men under arms in the course of the conflict. If perhaps 3.8 million adult males in Britain and Ireland were eligible for military service in this period (assuming that there were three million in 1775, and adding 3% per year for the 17-year olds entering the pool of theoretically available men), we have a military participation ratio of between one in seven and one in eight. This is less than the one in four (or five or six) estimated for the Napoleonic War, when invasion again threatened – let alone the nearly one in two calculated for the First World War. But it exceeds the ratios for earlier eighteenth-century conflicts (perhaps one in nine or ten for the Seven Years' War and only about one in 16 for the War of Austrian Succession). Indeed, one in seven or eight represents a significant mobilization of manpower, and casts serious doubt on the argument that for Britain the War of Independence was an old-fashioned, 'limited' conflict.

We still have a difficulty, however. A nation in arms, in the French Revolutionary sense, needs to be at least partly inspired by political belief. The Irish volunteers, as we shall see in Chapter 8, were a politically-motivated force, though in their case pro-American sympathies made them a decidedly doubtful addition to British strength. It might be assumed that most of the British sailors, soldiers, militiamen and volunteers who served in the American war lacked any ideological commitment. Surely compulsion, economic necessity, a desire for escape or adventure, and pressure from social superiors explain the vast majority of enlistments? When the Earl of Bute was urging the formation of a company of fenciblemen in the Western Isles in 1778, he appealed to tradition and locality. Mountstuart House, the Earl's ancestral home, was said to be under threat from raiding American privateers; if the threat materialized there could be no doubt that 'the Bute people to a man would take up arms'.[53] Bute was placing his faith in clan loyalty, not anti-revolutionary sentiment or patriotism.

But we should not be too hasty in dismissing the Britons under arms as politically remote from the issues of the war. We have already seen that there was a patriotic surge in response to the French entry into the conflict. We should also recognize that there were British officers – and ordinary members of the army and navy – who regarded their American opponents with intense hostility. Ensign Daniel Gwynne of the Ninth Foot told his brother that to obtain a proper idea of one of the colonists he should read *Hudibras*, Samuel Butler's seventeenth-century satire on the Puritans, and 'add to that as much Knavery, Roguery low Cunning and all kinds [of] Villany under the Mask of religion ... as you can well put together'.[54] Gynne's judgement was no doubt coloured by his being held captive by the Americans for four years; but many expressions of dislike seem to have had an ideological edge. Passages in British letters and diaries reveal a genuine – even a passionate – opposition to the 'sin' of rebellion. In 1775 Captain Walter Sloane Laurie of the Forty-third Foot railed against the 'seditious principles' of Joseph Warren, the leading Massachusetts patriot.[55] Thomas Sullivan, a private in the Forty-ninth Foot, recorded after the battle of Bunker Hill that the British troops were 'inveterate against the Rebels, on account of their ambitious designs'.[56] A year later, the wife of soldier in the Forty-seventh berated American troops for taking up arms 'agst your King & Countrymen';[57] while in the last days of 1776 Captain William Leslie of the Seventeenth Foot concluded that the inhabitants of New Jersey deserved the destruction and desolation of their province 'as much as any set of people who rebelled against their Sovereign'.[58]

This, of course, is to concentrate on attitudes within the armed services. Public opinion in Britain itself was divided over the war – or at least over war with the Americans. Indeed, the polarization of this period seems in many ways to prefigure the antagonism between pro-and anti-war factions of the 1790s, when the country was riven over British participation in the war against revolutionary France. A staunchly loyal Anglican clergyman wrote on 1 January 1777 of the American rebellion as an 'unnatural Proceeding'. He looked forward to the British army 'bringing back these ungratefully rebellious Children to their Duty'.[59] Other Britons were less enthusiastic. The Whig grandee Earl Spencer referred regretfully to the 'unfortunate contest with our American fellow subjects';[60] while a Worcestershire manufacturer declared: 'The generous & impartial, are I think in favour of the Americans in this country.'[61] The language employed, as we shall see, was not always so restrained; government supporters regularly accused their opponents of treasonably sustaining the rebellious colonists, and supporters of the Americans replied, just as bitterly, that the government was bent on establishing an authoritarian regime – both in the colonies and at home. And if the American war made its mark on British politics, it did not leave the British economy and British society unaffected – especially from 1778, when French, then Spanish, and finally Dutch intervention stretched to breaking point the resources available to George III's ministers. The immediate impact of the struggle on the mother country will be explored later, but for the moment suffice it to say that, even for Britain, the War of Independence was no 'limited' war.

Chapter 2 Notes

1. Quoted in Geoffrey Best, *War and Society in Revolutionary Europe, 1770–1870* (London, 1982), p.81.
2. See, for instance, Myron P. Gutmann, *War and Rural Life in the Early Modern Low Countries (*Princeton, N.J., 1980); John Gooch, *Armies in Europe* (London, 1980); John Childs, *Armies and Warfare in Europe, 1648–1789* (Manchester, 1982); M.S. Anderson, *War and Society in Europe of the Old Regime, 1618–1789* (London, 1989).
3. See Fred Anderson, *A People's Army: Massachusetts Soldiers and Society in the Seven Years' War* (Chapel Hill, N.C., 1984).
4. See David D. Bien, 'The Army in the French Enlightenment: Reform, Reaction and Revolution', *Past & Present*, no. 85 (1979), 68–98.
5. Cunninghame to Lord Barrington, 19 May 1776, WO 1/991.
6. Quarter Sessions Minute-book, 1778–81, QS 39/8, Michaelmas Session, 1778, Warwickshire RO.
7. Quarter Sessions Book 1777–82, Q/SMg 23, 14 July 1778, Essex RO.
8. *Writings of Washington*, iii. 433.

9. *Hamilton Papers*, ii. 18.
10. *The Revolutionary Journal of Baron Ludwig von Closen*, trans. and ed. Evelyn M. Acomb (Chapel Hill, N.C., 1958), p.89.
11. 'Military Journal Kept in 1777 ... by John Goodwin', Essex Institute, *Historical Collections*, xlv (1909), 207.
12. *Valley Forge Orderly Book of General George Weedon* (New York, 1971), p.213.
13. 'Diary of Enos Hitchcock', ed. William B. Weedon, Rhode Island Historical Society, *Publications*, new series, vii (1899), 223.
14. *The Lost War: Letters from British Officers during the American Revolution*, ed. Marion Balderstone and David Syrett (New York, 1975), p.180.
15. Alexander, Lord Lindsay, *Lives of the Lindsays* (4 vols., Wigan, 1840), iii. 225–6.
16. Percy to Gen. Sir William Howe, 9 Feb. 1777, Percy Papers, vol. xlix, pt. A, p.13, Alnwick Castle.
17. ... to Dr. Rogers, 23 April 1775, Rockingham MSS, R 150–2, Sheffield Central Library.
18. Capt. Baldwin Leighton to William Congreve, 20 April 1777, Congreve Papers, D 1057/M/F/32/1, Staffordshire RO.
19. Maj.-Gen. William Phillips to the Duke of Newcastle, 10 July 1777, Newcastle of Clumber MSS, NeC 2810, Nottingham University Library.
20. Memo by Maj. John André, n.d., but filed under May 1780, Clinton Papers, William L. Clements Library.
21. 11 Feb. 1781, Beverley Letter-book, 1761–93, Library of Congress.
22. 'Diary of a French Officer', *Magazine of American History*, vi (1880), 209.
23. Allan Kulikoff, *The Agrarian Origins of American Capitalism* (Chalottesville, Va., and London, 1992), p.163.
24. *The Papers of William Livingston*, ed. Carl E. Prince, *et al.* (5 vols., Trenton and New Brunswick, N.J., 1979–88), i. 135.
25. Gage to Barrington, 26 June 1775, Barrington Papers, H.A. 174/1026/6a (2), East Suffolk RO.
26. *Letter Book of the Rev. Henry Caner*, ed. K.W. Cameron (Hartford, Conn., 1972), p.165.
27. For a fuller investigation of the intimate connection between religious and political ideas see J.C.D. Clark, *The Language of Liberty 1660–1832* (Cambridge, 1994).
28. Quoted in John M. Palmer, *General von Steuben* (New Haven, Conn., 1937), p.157.
29. *Adams Family Correspondence*, ed. Lyman C. Butterfield (8 vols. to date, Cambridge, Mass., 1963–), ii. 328.
30. *Valley Forge Orderly Book*, pp.31–2, 168.
31. Connecticut Historical Society, *Collections*, vii (1899), 241.
32. Tirrell to Garrett Minor, 30 Sept. 1781, Garrett Minor Papers, Library of Congress.
33. Webb to Joseph Barrell, 11 July 1779, *Correspondence and Journals of Samuel Blachley Webb*, ed. W.C. Ford (3 vols., New York, 1893–4), ii. 180.
34. *General John Glover's Letterbook 1776–1777*, ed. Russell W. Knight (Salem, Mass., 1976), p.38.

35. *The Papers of Henry Laurens*, ed. Philip M. Hamer *et al* (12 vols. to date, Columbia, S.C., 1968–), x. 48.
36. Morris to John Jay, 10 Jan. 1784, *John Jay: The Winning of the Peace*, ed. Richard B. Morris (New York, 1980), p.674.
37. Howe to Lord George Germain, 7 July 1776, *DAR*, xii. 158.
38. Memo. of conversation, 7 Feb. 1776, Clinton Papers.
39. Rooke to Charles Rooke, 6 Jan. 1777, Rooke Papers, F3/13/1, Gloucestershire RO.
40. Hanbury Parish Register, D 1528/1/4, Staffordshire RO.
41. Money to Charles Jenkinson, 13 March 1779, WO 1/1004.
42. WO 34/225, fo. 267.
43. Sir Stanier Porten to Jenkinson, 9 March 1779, State Papers Domestic, SP 41/28, PRO.
44. Maj. James Barker to Matthew Lewis, 6 April 1781, WO 1/1010; Maj. George Reynolds to ..., 24 May 1779, WO 1/1005.
45. Lord Camden to Lady Chatham, 17 Nov. 1779, Chatham Papers, PRO 30/8/25, fo. 206.
46. Lt. Bennet Cuthbertson to Lord Althorp, 13 Oct. 1782, Althorp Papers, G3, British Library.
47. Thomas Babb to Christopher Hamlyn, 28 July 1778, 1292 M/Shipping, Devon RO.
48. Robert Rauthmell to his aunt, 6 July 1778, Rauthmell Letters, WD/AG/Box 3, Cumbria RO.
49. 'Mary Hardy's Diary', ed. B. Cozens-Hardy, Norfolk Record Society, *Publications*, xxxvii (1968), 43.
50. Lt. Matthew Bentham to the Earl of Ailesbury, 10 Oct. 1779, Savernake MSS, 9, Wiltshire RO. The writer appears to have been referring to volunteers.
51. James Wodrow to Samuel Kenrick, 23 May 1782, Wodrow-Kenrick Correspondence, MS 24.157 (74), Dr Williams's Library.
52. Lt. James Home to William Home, 10 Dec. 1782, Logan Home of Edrom Muniments, GD 1/384/7/19, SRO.
53. *Papers on Peter May, Land Surveyor, 1749–1793*, ed. Ian H. Adams (Edinburgh: for the Scottish History Society, 1979), p.206.
54. Gwynne to William Gwynne, 12 Oct. 1781, Gwynne Letters, D/CT/271, Pembrokeshire RO.
55. Laurie to [John Roebuck], 23 June 1775, Bannerman MSS, University of Glasgow Library.
56. Sullivan's Journal, p.29, American Philosophical Society.
57. Lt. & Capt. Sir Francis Carr Clerke to Lord Polwarth, 13 July 1776, Polwarth Papers, L 30/12/17/3, Bedfordshire RO.
58. Leslie to Countess of Leven and Melville, 25 Dec. 1776, Leven and Melville Muniments, GD 26/9/513, SRO.
59. Hanbury Parish Register, D 1528/1/4, Staffordshire RO.
60. Spencer to Lord Althorp, [28 May 1777,] Althorp Papers, G2.
61. Kenrick to Wodrow, 19 July 1775, Wodrow-Kenrick Correspondence, MS 24.157 (51).

3

The Protagonists

Before we return to a narrative of events, we need to know what the various participants in the war hoped to achieve, and how capable they were of fulfilling their objectives. In this chapter we shall examine the war aims and the strengths and weaknesses of the British, the Americans, the French, the Spanish, and the Dutch, looking in each case at the size and quality of their armed forces, the financial resources available to support their respective war efforts, some of the advantages they possessed, and some of the difficulties they faced.

*

Lord North and his ministerial colleagues had not prepared for a full-scale war in America; they had expected a show of force to subdue the rebellious people of Massachusetts. They had been taken by surprise when royal government collapsed in other colonies in the winter of 1774–5. What had started as a police operation became a major military effort. No one in government circles thought that the colonies could be permitted to go their own way. Other parts of the empire might try to follow suit, and America itself was a useful source of raw materials and re-exports and particularly valued as a major market for British manufactures: in 1772–3 the North American colonies accounted for 37% of all British exports. In part, then, the British were fighting to keep the 13 colonies in the commercial system created by the Navigation Acts – a system that was widely believed to underpin national prosperity and power. But there was also an ideological aspect to British government thinking. This was a war against unlawful rebellion; a war to restore legitimate authority in the colonies. There was much talk of rescuing loyal Americans from the 'tyranny' of Congress.

The entry of the French into the war, then the Spanish, and finally the Dutch, changed the nature of the conflict and with it Britain's war aims. North's government still hoped to keep the Americans within the empire and Britain's commercial orbit – to this end in 1778 Parliament even gave up its claims to the right to tax the colonies – but now ministerial attention was re-directed by the requirements of what can be described as a world war. The new strategic focus for the British was the Caribbean, where their islands were seen as at least as valuable as the 13 colonies, and the government also felt compelled to protect Britain itself, the Mediterreanean outposts of Gibraltar and Minorca, and British possessions in Africa and India. But the British were not thrown entirely on the defensive. They launched offensive operations in the West Indies, and attacked Spanish territory in Central America. Ambitious plans were drawn up to take the war to the Spanish settlements in South America and the Pacific (it was hoped that the Spanish Pacific territories would provide spices and gold for exchange in the China trade of the East India Company). In India, too, the British did not timidly wait for their enemies to strike. When news arrived of the French war, the East India Company's forces rapidly overran the French commercial posts. In 1781, when reports of the Dutch war reached India, the British treated the stations of their Dutch rivals in the same manner. The Dutch, indeed, were seen by the British as an easy target elsewhere. Many of their slave-trading posts in West Africa were seized, an attempt was made on the Dutch colony at the Cape of Good Hope, Dutch bases in Sumatra were taken, and consideration was given to the capture of other Dutch outposts in the East Indies. British ministers, in short, while fighting against a powerful coalition, still hoped to expand the country's power and prosperity.

What were the means at their disposal? When the fighting broke out at Lexington and Concord, the British army's paper strength was 48,647 officers and men. In reality, it could probably muster no more than 36,000. These troops were scattered widely over the areas of British involvement: in Britain itself, in Ireland, in Gibraltar, Minorca, Africa, the West Indies; only around 8,000 were in the mainland North American colonies. In India the interests of the East India Company – a territorial power as well as a commercial concern – were mainly protected by its own private army, for the most part made up of locally-raised Indian troops, which in 1771 numbered some 64,000 officers and men. The Royal Navy, the premier branch of Britain's armed forces and the one that absorbed most of the military budget, had an establishment strength of 16,000 seamen and in 1776 possessed some 340 vessels of all kinds.

At first, the expansion of the armed forces was decidedly slow. Lord North and many of his colleagues were reluctant to provoke the French by a full-scale naval mobilization, and it was not until October 1776 that reports of French fleet augmentation persuaded the cabinet to heed the advice of the admiralty and start putting the navy on a war footing.[1] The army recruited more men but put them into the existing regiments rather

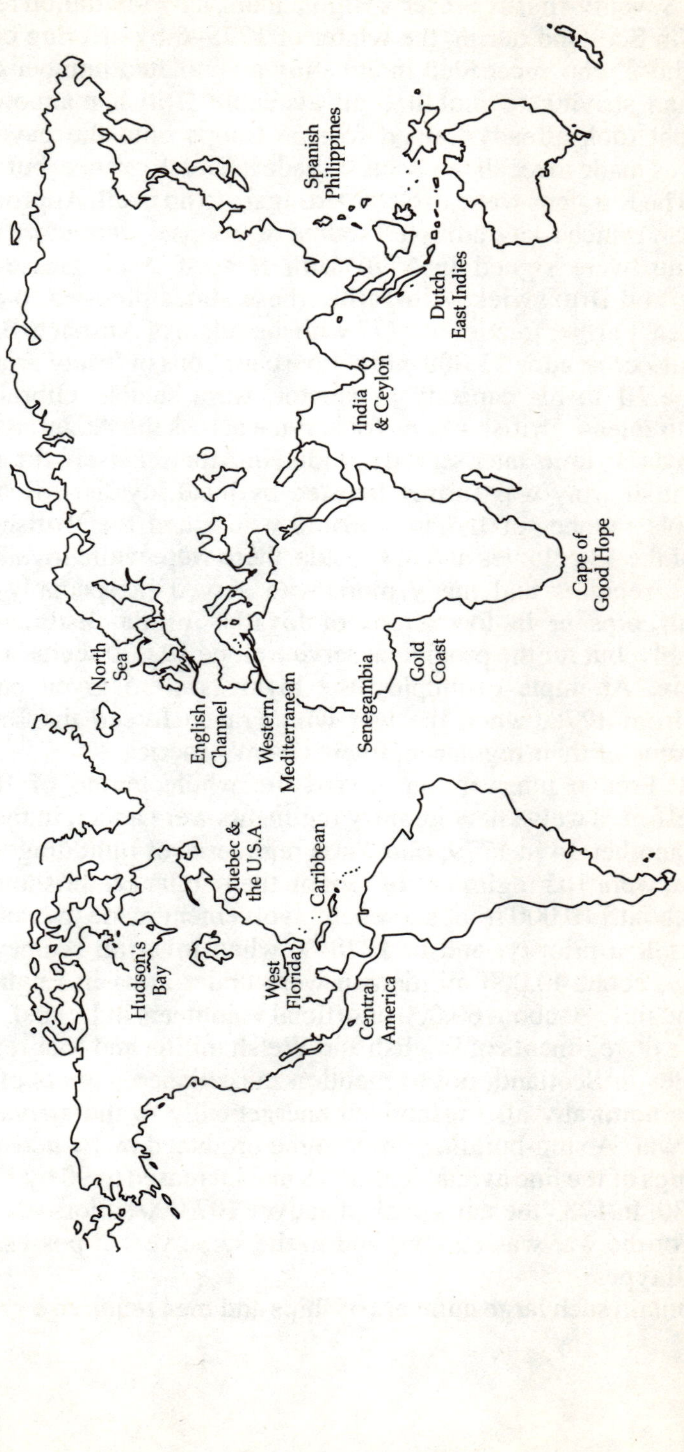
Hudson's Bay
West Florida
Central America
Quebec & the U.S.A.
Caribbean
English Channel
North Sea
Western Mediterranean
Senegambia
Gold Coast
Cape of Good Hope
India & Ceylon
Dutch East Indies
Spanish Philippines

A world war

than raised new ones. Only one new corps was formed between 1775 and 1778: the Seventy-first or Fraser's Highlanders, a two-battalion regiment recruited in Scotland during the winter of 1775–6 by offering commissions to those who succeeded in enlisting a stipulated number of men. Rather than striving to mobilize all available British manpower, the government took already-trained foreign troops onto the payroll. An attempt was made to recall the Scots brigade in Dutch service, but without success. The Russians were also courted, again to no avail. Approaches to the German princes – a traditional source of troops – were more fruitful. Agreements were signed in 1776 with Hesse-Cassel, Hesse-Hanau, Waldeck, and Brunswick. Altogether these states pledged to provide 18,000 men. Further treaties in 1777 with the rulers of Anspach-Bayreuth and Anhalt Zerbst added 3,000 more. Five battalions of Hanoverians, lent by George III in his capacity as elector, were sent to Gibraltar and Minorca to release British troops for service across the Atlantic; another two Hanoverian units later served in India. In America itself, the strength of the British army was further boosted by local loyalists. Besides an incalculable number of Indian warriors who aided the British on the frontier of the 13 colonies and in Canada, there were white loyalists who joined the regulars and many more who served in specially-created provincial corps or in formations of loyalist militia. Estimates vary considerably, but for the provincial service alone 19,000 seems a reasonable figure. Attempts to employ the loyalists were given particular urgency from 1778, when the war with France forced the British to remove some of their regular regiments from America.

Indeed, French intervention altered the whole tempo of Britain's military effort. Twelve new infantry regiments were added in the spring of 1778; another 14 in 1779, plus three regiments of light dragoons. By 1781 there were 105 regiments of foot on the regular establishment, and all in all about 110,000 troops. French involvement made the defence of Britain itself a priority, and in 1778–9, when invasion seemed a real possibility, about 40,000 militiamen were under arms in England and Wales, and in 1780 about 60,000 unofficial volunteers in Ireland. In 1781 there were 66 regiments of English and Welsh militia and four regiments of fencibles in Scotland, not to mention miscellaneous units of volunteers. The admiralty also responded energetically to the arrival of the Bourbon war. A ship-building programme produced many new vessels. The 66 ships of the line available in 1778 had increased to 90 by 1779 and 95 by 1780. In 1783 the navy peaked at over 107,000 sailors (the annual average for the war was 82,000) and in the same year it possessed 617 craft of all types.

To maintain such large numbers of ships and men required a great deal of money. The East India Company was able to use the territorial revenues of Bengal, Bihar and Orissa, granted by the Mughal emperor in 1765; but the Company was already in financial difficulties before the war began, and by the end of 1782 one of its officials at Calcutta was lamenting the 'Immense Sums sent from hence to Bombay & Madras' to fund the Company's armies, and he concluded thoughtfully that for a commercial organization 'War is a dear Hobby Horse and Soldiers expensive Play things'.[2] It was costly, too, for the British government, which was obliged to provide for the expansion of the forces in its own pay. During the American war the army, navy and ordnance cost more than £109 million (compared with nearly £83 million in 1756–63 and just under £56 million in 1739–48). In 1780 military spending was the equivalent of 12.5 % of estimated National Income; a larger proportion than in the French Revolutionary War, when Britain is generally thought to have been placed under immense strain (in 1800 military spending accounted for 10.4 % of a much-increased National Income).

How were such vast sums raised? By 1775 Britain had a tried and tested system of war finance. In the 1690s England experienced a revolution in her fiscal and credit arrangements that followed from, and was clearly related to, the constitutional changes of 1688–9. The position acquired by Parliament after the Glorious Revolution – it became a permanent institution rather than an occasional event – and the willingness of the new monarchs to work with and through Parliament instead of against it, enabled governments to raise taxes more efficiently and to take out long-term loans on the security of regular parliamentary imposts. The 'fiscal-military state', as John Brewer calls it,[3] was firmly in place by the start of the Seven Years' War. During the American conflict, this well-honed system allowed £12 million a year to be raised in taxation; about 20% from the land tax (which rose to four shillings in the pound) and the rest from indirect taxes, mainly customs and excise. Much of this revenue was used to service interest payments on the national debt, which was thereby able to rise from £127 million in 1775 to £232 million in 1783. Almost half the cost of the war was in fact covered by borrowing.

But smoothly as the financial machinery seemed to operate, the unprecedented demands of the American war stretched it to the limit. By the end of the struggle loans could be raised only by offering substantial inducements in the form of free annuities: the government's credit-worthiness was very much in doubt. And the war-time tax burden – the share of per capita income appropriated in taxes reached 23% – brought forth cries of anguish. In May 1780 even government supporters were reported to be 'growl[ing] most confoundedly at the increasing load of taxes wch we all labour under':[4] indeed, this mood of tax mutiny contributed significantly to the country-wide political protests that so unsettled Lord North's ministry in 1779–80.

Of more immediate concern to the army and navy commanders was the quality of the forces at their disposal. They were certainly more professional – at least in the early stages of the war - than their American opponents. Some British officers, recalling the American performance during the last war, were confident that colonial resistance would collapse when faced with a show of force from a sizeable body of British regulars. But we should resist the image of a near invincible British military machine favoured by historians whose aim – consciously or subconsciously – is to magnify the achievements of the amateurish Americans. The navy, seriously underfunded since the end of the Seven Years' War, was generally thought to have become less efficient; to be, indeed, a shadow of its former glorious self. Although the navy could still boast far more ships of the line than any of its competitors, dockyard facilities had fallen a victim to government spending cuts and many of its vessels were in a poor condition. The army, like the navy, had not been tested in action since 1763. During peace-time it was customary for regiments to function with only a minority of their officers present and serving; and adjusting to more demanding conditions seems to have caused some difficulty: in 1781, when the French attacked Jersey, the king was obliged to complain of the shockingly high rate of absenteeism. The quality of those who were present could also cause unease. Some of Gage's officers who were incapable of making the transition to active soldiering had to be edged out of the service. In 1775 Major Isaac Hamilton of the Royal Irish was apparently thought unfit to remain in such an important position: Lord Barrington, the secretary at war, wrote disapprovingly of Hamilton's 'inability', and a month later the major resigned his commission.[5] Nor were the ordinary soldiers always well-drilled veterans. About a quarter of the infantry was made up of recruits – men with less than a year's service. Even experienced troops were not necessarily properly trained. Regiments based in Ireland tended to be dispersed in small detachments to perform police duties in support of the civil authorities. Some units remained divided in this way for years at a time, making it impossible to organize effective regimental training. Before departing for America the Twenty-eighth Foot was said to be in a deplorable state. 'The men are slouching and ill set up', wrote their inspecting officer, 'They are not Steady and do not know how to handle their Arms'.[6]

The German auxiliaries also varied in quality. Most accounts agree that the first detachments to reach America in the summer of 1776 were made up, on the whole, of well-trained and formidable soldiers; the problem lay with the recruits sent out subsequently. Sir George Osborne, whose task it was to inspect the arriving Germans, noted in 1777 that the Waldeckers were 'a very low Size, very young and very bad recruits'.[7] Johann Ewald, a Hessian *Jäger* officer, similarly commented on the 'deserters and insolent rabble' sent out to bring his corps up to strength: 'blood turned to water and all spirits sank' among his fellow officers, 'who until now had commanded the most upright and obedient of men'.[8]

The American loyalists, of course, were no more experienced in military matters than were their revolutionary countrymen. A few ex-regulars served in the provincial line, but they were not usually numerous enough to counterbalance the military deficiencies of their colleagues. 'Tho' our Commissioned Officers are good Spirited Gentlemen,' Major James Gray of the King's Royal Regiment of New York wrote in 1779, 'yet they are equally ignorant of Military discipline, never having been in the service nor as yet had time to learn their duty'.[9] Some of the loyalist units eventually developed into respectable bodies of men, the equal of the British regular regiments; but other corps proved more of a liability than an asset. The Roman Catholic Volunteers were disbanded in 1778 owing to 'the inattention of the officers to the terms of their warrant and their utter disregard of all discipline'.[10] Emmerick's Chasseurs suffered the same fate in 1779.

Nor did the British militia initially inspire much confidence. Contemporary cartoons depicted them as a decidedly motley crew, with more than its fair share of the ill-clad, the physically unfit and the mentally retarded. But many Englishmen waxed lyrical about the qualities of the militia, seeing them as the embodiment of public virtue – a citizen soldiery inspired by love of country rather than the mercenary considerations of a standing army. In April 1778 Sir William Jones, an Oxford academic, congratulated Viscount Althorp on becoming a captain in the Northamptonshire militia and announced grandly: 'Neither the French nor any other enemies of England, can ever hurt us while our young men are animated by such a spirit'.[11] In time martial enthusiasm was matched by a degree of military efficiency. The militia were trained alongside regular troops in summer camps: from Coxheath in Kent an officer reported that they were 'constantly at exercise every morning and Evening'.[12] The militia were never tested in combat, so we can only speculate on how they might have performed against an invading army, but several regiments earned praise for their putting down of local riots: in 1783, for instance, the Staffordshire militia was commended for its 'steady and faithful conduct'.[13]

The forces of the East India Company, large as they were, caused some concern to the regular officers serving with them. Sir Eyre Coote, a hero of the Seven Years' War in India, noted in 1780 that the European-trained sepoys were no longer such an asset as they had been because the troops of the princes hostile to the British were 'so much improved in Discipline, and steadiness, since I was in India before, that they are too much on a par with ourselves'.[14] In many senses the small body of European soldiers was much less impressive, even though it was always seen as the reliable nucleus of the Company's army. Of the European recruits sent to India in 1779, 32% were under 16 years old. Worse still, a significant proportion of the Company's European troops seem to have been known or suspected criminals – a much more significant proportion than in the regular army. The Company's ranks included many men like William Neale, a highwayman convicted at the Maidstone assizes in March 1778 and pardoned a year later on condition that he joined and continued 'in their Service in the East Indies'.[15] Sending out regular troops to bolster the Company's forces was felt to be the only way to improve the situation, though Coote was to complain that the Highland regiments that arrived were ill-suited to the climate and prone to sickness.

Besides marked differences in the capability of the forces at their disposal, British commanders often had to contend with serious internal dissentions. In India Coote protested vigorously at the way in which the Madras council of the East India Company constantly interferred with his command, trying to deprive him of all power. After the battle of Ushant in July 1778 the Royal Navy was bitterly divided between supporters of Admiral Keppel and those who backed Admiral Palliser: each accused the other of misconduct in the battle, and the ill-feeling was increased by Keppel's close contacts with the parliamentary opposition, which did its best to exploit the situation to embarrass North's government.

This was probably the most public of the disputes within the British armed forces in the period, but it was by no means the only one and perhaps not the most damaging. In America General Clinton fell out with several of his army colleagues, and he regularly disagreed with Admiral Arbuthnot over political and operational matters, which can hardly have helped the British cause. Nor was it simply a case of unfortunate personality clashes. The army and navy were able to co-operate on occasion to carry out successful amphibious operations: in 1776 an American general noted enviously how 'the enemy (furnished with canvas wings) can fly from one spot to another',[16] and when the Howe brothers began the campaign at New York a British observer commented that 'we have every advantage that a perfect Cordiality between the Admiral and the General can give us'.[17] But for much of the time relations between the land and sea services were far from harmonious. Disputes over the distribution of prize money – derived from the sale of goods captured from the enemy –

caused particular friction. The wrangling after a raid on Virginia in 1780 dragged on literally for years, being finally resolved only in 1793. Small wonder one of Clinton's officers wrote in despair: 'I wish to God all Prises were immediately destroyed'.[18]

Divisions within the army were, if anything, even more acrimonious. Some British officers could manage a few good words about the Germans or the loyalist provincials – 'a fine Body of men', was how Captain Ralph Dundas of the Royal Navy described the Hessians[19] - but for the most part they viewed their auxiliaries with ill-disguised contempt. The Hessians were condemned for their allegedly slow manoeuvering and for systematic plundering of the inhabitants: 'beyond belief – ', wrote Charles O'Hara of the Guards, 'Cruel & Savage'.[20] British-loyalist relations – already strained by the disdain of the regulars for the inexperienced provincials – deteriorated still further in 1779 when a reform of regulations, designed to stimulate loyalist recruitment, allowed provincial officers to take precedence over British officers of lower rank. Clinton was presented with a memorial signed by many of the army's field officers protesting at the change and asking the king to reconsider. Not surprisingly, both the Germans and the loyalists responded sharply to what they saw as ill-treatment. Captain Johann Hinrichs of the *Jägers* complained of the 'confounded pride and arrogant bearing of the English',[21] while John Chalmers, commander of the Maryland Loyalists, observed coldly that 'equality of manners' in America 'renders such pride more odious than in any other country'.[22]

By using auxiliaries the British probably made the task of pacifying the colonies much more difficult. 'O Britons, how art you fallen that you hire foreigners to cut your children's throats', exclaimed a Rhode Island officer;[23] and it seems that many Americans were spurred on to more determined resistance by the presence of the Hessians in the British ranks. The Indians were even more doubtful allies. Their bloodthirsty conduct while serving with Burgoyne's army in 1777 appears to have brought many more Americans into the field than might otherwise have been the case: 'It would certainly have been better,' a Brunswick officer lamented, 'if we had not had any Indians with us.' [24] The loyalist provincials and militia, many of whom had lost property to the revolutionaries and were anxious to settle old scores, caused similar problems. No unit was more notorious than the Associated Loyalists, militant refugees who were unwilling to join the provincial line and submit to military discipline. Their name became a byword for depredation and theft. Clinton, according to a revolutionary outraged at the activities of the Associated Loyalists, was no more than 'the ringleader of a banditti of Robbers!'[25]

All this mattered because this was a very different kind of war from those that the British were used to fighting. Not since the Jacobite uprising of 1745 had they been faced with a remotely similar task. Suppressing rebellion in America was not the same as countering the French or Spanish in Europe or the colonies. The attitude of the local people acquired a special importance. British commanders were acutely aware of this unconventional dimension to the conflict: indeed, as we have seen, one school of thought suggests that constant attention to the political aspects of the war hampered effective prosecution of military operations.[26] But whatever army and navy commanders wished, the British forces did much to alienate loyal and potentially friendly Americans. Requisitioning of supplies, seizure of domestic and commercial property to accommodate the troops, and the harshness of martial law were just a few of the grievances expressed by those who resided in the occupied areas or within reach of British detachments.[27] And it was not only the Hessians and loyalist irregulars who plundered the local people: in June 1777 the commander of the Fortieth Foot decried the 'most Shameful practice of Maroding & plundering' that prevailed in the regiment;[28] while during the British retreat across New Jersey in June 1778, Lieutenant-Colonel Alured Clarke of the Royal Fusiliers similarly reprimanded his corps, telling the troops that he was 'Mortified at Observing the great Irregularity and excesses that have been Committed'.[29] As a percipient British commentator noted, 'Those who formerly wishd our approach, And would with Joy have seen Us triumph Over the rebels, will now Arm to defend their All from Undistinguished plunder'.[30] If the British commanders could not stop their men from abusing American civilians, their chances of winning new friends, or even keeping existing ones, were decidedly limited.

A major problem for the British was the size of the theatres of operation. Distances in the traditional European regions of conflict were relatively small – from Lille to Maastricht is only about 120 miles, and the Duke of Marlborough's famous 350-mile march from the Rhine to the Danube in 1704 was regarded as the wonder of the age. In India, by contrast, over 1,000 miles separated the Bombay and Calcutta bases of the East India Company, and to co-ordinate campaigns against the Marathas and Haidar Ali of Mysore was no mean feat. Similarly, fighting in America took place along about 1,200 miles of the Atlantic seaboard. An effective naval blockade, the mode of action preferred by some British politicians and military figures, would have been difficult to maintain with the limited number of vessels available (a significant proportion of the Royal Navy had to be kept in home waters, even before the French intervened, as a safeguard against invasion). Land campaigns were also rendered more problematic by the vast distances involved, especially when they were fought away from the coast or navigable rivers and in areas that were poorly charted and sparcely populated. The capture of major towns brought little advantage to the British: at one time or another they occupied Boston, New York, Philadelphia, Newport, Charleston and Savannah; yet the seizure of these population centres failed to break the rebellion. The possession of territory, as the British commanders in the field realized only too well, was less important than the defeat of the Continental army. But trapping and crushing the Continental army proved impossible for the British so long as Washington was prepared to concede territory in order to survive and fight another day.

Another geographical problem for the British was posed by the need to conduct operations so far from home. The army that sailed to the colonies in 1776 was the largest expeditionary force that the British had ever sent abroad. Once in America, it had to be sustained in a mainly hostile environment. It soon proved impossible to supply the army from local sources alone. Considerable quantities of provisions and hard cash therefore had to be shipped 3,000 miles across the Atlantic. At times this logistical problem was overcome in an impressive manner: 'The Supplies of Provisions from Europe have been so wisely planned and so well executed', Howe reported from New York towards the end of the 1776 campaign, that his army could 'be under no Apprehension of Want'.[31] But such an extended supply line was obviously exposed, especially after the French entered the war. As Clinton observed in November 1778, 'we have not at this time above 24 days bread or flower, I am sorry to say I am a little accustomed to this sort of neglect, 4 times since I came to this Command have I been within 3 weeks of Starving'.[32] By 1782, according to one authority, the war would have ground to a halt even without the demoralizing British defeat at Yorktown, for British shipping resources had finally been stretched beyond the limit. It would have been very difficult to mount any further offensive operations in America.[33]

In truth, the British were not just stretched logistically. Britain's success in the Seven Years' War was based on tying France down to a land war in Europe, conducted on Britain's behalf chiefly by subsidy allies, while the British were free to devote most of their military effort to a colonial and maritime war. The French, in short, had been obliged to face in several different directions at once, and they had little choice but to give priority to the defence of their interests in Europe. In the American war the tables were turned. It was the British who were committed to fighting on many fronts. They were the ones engaged in a major continental war – in America – while the French had almost no European distractions. In the absence of war in Europe (apart from operations at Gibraltar and Minorca in support of the Spanish, and preparations for an invasion of Britain) the French were able to concentrate on a maritime and colonial war in the West Indies and India. Over-extended, the British could hardly hope to hold their own in every theatre. As George III observed when in March 1781 news reached Britain from India of the defeat by Haidar Ali of Lieutenant-Colonel William Baillie's detachment the previous September: 'I do not see what can be done ... we certainly have neither troops nor ships that can in the present situation be taken either from hence, North America, or the West Indies'.[34] Given the odds against them, the British were perhaps fortunate to avoid even more serious defeat. But this assessment, of course, ignores the considerable problems confronted by Britain's enemies.

*

The Americans consistently saw themselves as parrying British aggression; they even maintained that this defensive stance was compatible with an invasion of Canada in 1775 and later plans to repeat the forcible union of all the North American colonies. But American war aims, like those of the British, changed as the conflict developed. If 1778 was the year of reappraisal for the British, for the Americans it was 1776. For the first 14 months or so of the struggle the American objective was to resist British military coercion in order to oblige North's government to back down and accept a return to something like the imperial *status quo* of 1763. Initially, in other words, the aim was to remain within the empire, albeit an empire in which the British Parliament played a much less important role. But the events of the war, as we shall see later, altered the American agenda. From July 1776, rather than trying to defend their own vision of empire, the Americans were trying to sustain their independence from Britain.

Their prospects of success did not at first look bright. At the beginning of the war the rebellious colonies had no navy and no army; they had only a determined, enthusiastic but poorly-trained militia. Congress voted to create a navy in the autumn of 1775, when it authorized the construction of five warships. The embryonic Continental navy had its successes – intercepting British supply vessels and raiding Nova Scotia; and John Paul Jones, the navy's most celebrated officer, boosted its reputation by taking the war into British waters. But the Continental navy suffered a serious reverse when it allowed itself to be bottled up at Providence, Rhode Island, in December 1776. More dramatically, it lost vital ships in the Penobscot fiasco of 1779 and when the British took Charleston in 1780. Indeed, for the most part the American contribution to the war on the ocean came not from the regular navy but from a host of privateers. In the first year of the war alone, these seaborne irregulars seized 55 British vessels. But none of this could disguise the fact that until the French entered the war the Americans were at a marked disadvantage in naval terms; and once the French navy was available, any incentive to spend scarce money on the development of an American navy disappeared.

On land the colonists were numerically more formidable. Great numbers of Americans – precisely how many remains unclear – were called out at one time or another as militiamen during the conflict, either to defend their own communities or to reinforce the Continental army. This army, established in June 1775 when Congress took control of the 17,000 New England troops besieging Boston, was soon to become the life and soul of the Revolution, or as General Nathanael Greene put it, 'The Stamina of American liberty'.[35] In the dark days of September 1776, Congress authorized the raising of an army of about 75,000 officers and men. At no time, however, was the Continental army more than 34,000 strong. It was made up of successive relays of recruits drawn from the militia who remained with the colours for anything from a few weeks to three years. As a result, by the end of the war the Continental army had employed the services of a good many Americans.

The quality of the revolutionary troops varied enormously. The Continental army, as Washington recognized, was formed under circumstances that made unlikely the early attainment of 'the order, regularity, and discipline of veterans'. But there were, in Washington's view, special difficulties that arose from the democratic practices of the New Englanders: 'the officers of the Massachusetts *part* of the Army', he complained in August 1775[36]

> are *nearly* of the same kidney with the Privates ... there is no such thing as getting of officers of this stamp to exert themselves in carrying orders into execution – to curry favor with the men (by whom they were chosen, and on whose smiles possibly they may think they may again rely) seems to be one of the principal objects of their attention.

The different periods of service also hampered the creation of an efficient fighting force. 'Had the army at first been enlisted for the [duration of the] war, what a happy circumstance would it have been', William Heath of Massachusetts reflected ruefully in July 1776. 'A six weeks, two months, &c. militia has prevented our having a proper disciplined army.' [37] More than four years later a Connecticut officer was voicing the same concern: 'We all wish the New England states would seriously attempt to raise an Army – to raise troops for the continuance of [the] war – or some long period – and give over trifling with their six-months-men'.[38]

In time, despite these disadvantages, Washington and his colleagues were able to remodel the Continental army, imposing upon it greater discipline and uniformity. The comments of its opponents are instructive here. General James Grant, who held the Americans in special contempt, was at first inclined to attribute their (to him unexpected) military success to the former British officers and Irish and Scots immigrants in their ranks. American-born soldiers, he argued, were incapable of military ardour. Sentiments like these continued to be expressed throughout the war, but as the conflict dragged on a new tone of respect can be detected. Even Grant changed his tune, modifying his stance in a parliamentary speech in 1779. Other British officers were more generous. In October 1777, after hearing of the defeat of General Burgoyne at Saratoga, Ensign Lord Dunglass of the Guards wrote: 'The American Rebels are by no means so despicable an Enemy as they are thought to be in Britain ... they are every day acquiring Military Knowledge and Improving in Discipline'.[39] The following December Captain Ewald observed that the Americans, by assiduosly attending to military literature, had produced 'many excellent officers who very often shame and excell our experienced officers, who consider it sinful to read a book or to think of learning anything'.[40] During that same winter of 1777–8 Baron von Steuben subjected the American troops to rigorous training and reorganization. Revitalized, they went into the field for the 1778 campaign in perhaps a better state of discipline and efficiency that at any time in the war.

The American militia was less impressive as a fighting force. True, it had its moments. At Bennington, in the Burgoyne campaign of 1777, New Hampshire militiamen humbled German regular troops; and immediately after the British retreat from Lexington and Concord Earl Percy warned that the New England militia should not be underestimated: 'they have men amongst them who know very well what they are about'.[41] But all too often the performance of the militia disappointed American commanders. Colonel Daniel Hitchcock of the Eleventh Continental Infantry complained in September 1776 that 'The Militia are not worth a farthing'.[42] At Rhode Island in 1778 General John Sullivan moaned at 'the unsoldierlike Behaviour of the Militia who had engaged to serve for fifteen days, & who could not have been prevailed upon to remain One hour after the expiration of that time'.[43] Reluctance to serve for prolonged periods was compounded by irresolution under fire: after the American defeat at Camden, South Carolina, in August 1780, a Continental officer let rip at 'The infamous Cowardice of the Militia of Virginia and North Carolina', who fled on the first sight of the British.[44]

Cooler reflection might have suggested that the deficiencies of these part-time soldiers were only to be expected. It was unrealistic to believe that they could match their more professional friends or enemies. Perhaps more surprising was the presence in South Carolina of militiamen from

North Carolina, let alone from Virginia. Particularism, or provincial loyalty, undermined the American war effort. The British threat might have helped to forge a sense of national identity, yet each of the states jealously guarded its own interests. They raised troops for home defence but were reluctant to release them for service with the Continentals. The mobility of the British – greatly increased of course by the strength of the Royal Navy – only added to this caution: no state wanted to find itself unable to resist a British incursion because it had committed too many of its men elsewhere. To overcome this way of thinking was no easy matter. Nathanael Greene, pleading with the governor of Delaware for troops to use in the South, mustered all his persuasive powers: 'Every wise People will keep the War as far from them as possible. The Middle States have no Way of effecting this but by giving timely Support to the Southern Operations'.[45] Fine words, but they hardly answered the objection that the British might just as easily come from New York to the north as from South Carolina, and that they were most likely to descend on the coast of the middle states. In private, Greene despaired: 'The local policy of almost all the States is directly opposed to the great National plan; and if they continue to persevere in it, God knows what the consequences will be'.[46]

Serious shortages of equipment, clothing, food and money also limited the military potential of the American forces. Weapons were widely distributed in the colonies, but even after 23,000 French muskets had been imported in 1777 American commanders regularly reported difficulties in arming all the available troops. As the marquis de Lafayette observed while serving in Virginia in the spring of 1781, 'Shou'd it be possible to get arms some more militia might be brought into the field'.[47] Uniforms were scarce, too. Supplies from France enabled the Continentals to appear in new blue or brown coats for the 1778 campaign, and captured British uniforms were recycled. Yet even after the introduction of the first dress regulations in October 1779, American troops were routinely described as 'without Clothing', 'almost naked', and so on.[48] Food shortages were still more alarming. At Valley Forge in the winter of 1777–8, Washington's depleted Continental army survived on firecakes, a thin bread made of flour and water heated over the campfire. On other occasions American troops were obliged to requisition provisions, or turned to the even more desperate expedient of theft. As Greene noted in December 1780: 'to have a good army you must begin by providing well for the Belly .... if an Army is not well provided, the People will soon begin to feel the Hand of violence'.[49] A lack of cash caused further difficulties; the mutinies in the Pennsylvania and New Jersey lines in January 1781 were primarily the result of prolonged failure to pay the wages of the soldiers.

Many of these logistical problems arose from the method of war finance forced on Congress. Congress itself could not levy taxes – Americans who had resisted parliamentary taxation were loathe to grant taxing powers to their own new all-American legislature – and the individual states, though they taxed their people heavily to raise significant sums for their own purposes, provided very little for congressional use. Before 1780 only $2.4 million in state taxes were contributed to the general war effort. Without an adequate tax base Congress could not hope to emulate the British system of deficit financing. A small tax revenue meant no guarantee of interest payments on loans, which (when combined with lack of confidence in ultimate American victory) effectively ruled out large-scale public borrowing. Congress, and indeed many of the states, therefore resorted to printing money. By the end of 1775 Congress had issued $6 million in paper bills. In 1776 another $19 million followed, $13 million in 1777, $63 million in 1778, and $140 million in 1779. By 1780 over $240 million had been issued in congressional bills and another $200 million by the individual states.

Some historians see this willingness to abandon conventional financial arrangements – a revolutionary financial policy to fund a revolutionary war – as a source of strength.[50] In theory, the financial weakness of the Americans should have forced them to submit; in reality it forced them to be bold and to cut free from the normal restraints. The result, in this optimistic interpretation, was a much more rapid and complete mobilization of American resources than would otherwise have been possible. But one could argue that the Americans overdid it. By printing such large sums they almost destroyed their own capacity to continue the war. As more money poured from the presses, its value naturally plumeted. As the Continental currency depreciated, Americans became increasingly reluctant to sell supplies to their own army. Even the recall of the paper issues from 1780, and an attempt to return to hard money, failed to restore confidence. The damage had been done. Many Americans saw the French army as a much more attractive proposition, and tried to sell their goods and services to it instead of the Continental forces. When the French expeditionary force marched south in 1781 to trap the British in Virginia, Lieutenant comte de Clermont-Crèvecoeur of the Auxonne regiment of Artillery noticed how American civilians 'crowded round ... loaded with every sort of produce, so that the camp was a continual market'.[51] In Virginia itself, George Mason, a fervent patriot, succumbed to the same lure. Having lost several stacks of hay to the Continentals ('for which I suppose I shall receive Certificates to be paid, when or how, I know not') he thought it 'but reasonable' to 'sell the remainder to the French Troops for Specie'.[51] Worse still, many more people preferred to exchange their commodities for British hard cash.

Once significant numbers of Americans started to trade with the enemy, further problems followed. They increasingly swapped their goods not for British money but for British manufactures, which had been officially boycotted by the colonists since the formation of the Continental Association in 1774. Such was the craving for these long-denied manufactures that many Americans, having traded in their own produce, even used their limited supplies of precious hard cash. By the last years of the war it was said that in some states this illegal commerce was drawing off nearly all the specie in circulation, making it almost impossible to raise local taxes to support the struggle. The Americans were thus caught in a downward spiral leading to military ineffectiveness. Their ability to sustain the Continental army had been undermined by their very system of war finance – or, more accurately, by their printing of massive quantities of paper money until 1780, the repercussions of which continued to be felt until the end of the war. It was probably only considerable subsidies and loans from the French, the Spanish and the Dutch (worth in all more than $10 million) that prevented total military collapse.

*

French government policy before and during the American war was profoundly influenced by the humiliation of defeat in the Seven Years' War. The lesson for internal reformers like Turgot, controller-general of finances when the first shots were fired at Lexington and Concord, was that France should steer clear of another war and concentrate on domestic reconstruction. But for other ministers, France's disastrous showing in the Seven Years' War pointed above all to the need for external not internal change; Britain, the chief beneficiary of France's misfortunes and her main commercial and imperial rival, had to be humbled. In 1776 the advocates of an aggressive foreign policy had their way, Turgot was defeated, and France started to aid the Americans. Two years later France officially became a belligerent. To Vergennes, the foreign minister, and others who thought like him, the American rebellion had to be sustained by France because the loss of the 13 colonies would weaken Britain economically, and would therefore reduce her status in international politics. France might also gain more directly if she were to supplant Britain as America's main trading partner. And there was the consideration that, while the British were occupied in North America, France could seize the chance to improve her position in the West Indies: the Caribbean islands were as highly valued by the French as by the British. But other areas of the globe attracted French attention as well. French interests could be advanced in Africa – a source of slaves for the West Indies – and India, where French trade more than doubled between 1769 and 1776. Moreover, the commitment of troops and ships to these theatres (and to European operations) had the advantage of diverting British resources from America and the Caribbean. Once the Dutch entered the war, the defence of their trading stations and colonies became another French objective; for the power of the British would be immeasurably enhanced if they were allowed to add the Dutch commercial empire to their own. The most obvious fruit of this aspect of French policy was the despatch of French forces to sustain the Dutch colony at the Cape of Good Hope; but French operations in India were at least partly influenced by the desire to tie down the British on the subcontinent to prevent them from encroaching on the Dutch outposts in the East Indies.

Despite her defeat in the Seven Years' War, France was unquestionably a great power; the power that most Britons continued to see as a serious threat and as a natural enemy. France's potential could not be denied. Her population, at around 25 millions, was about double that of Great Britain and Ireland and perhaps ten times as large as that of the 13 colonies. Not surprisingly, therefore, the military strength of France – at least on paper – was truly awesome. Her peace-time standing army numbered more than 150,000 officers and men; and around the same number of militiamen were available for home defence. But only a fraction of this vast array was actively employed during the war. In India, for instance, the British had to contend not with large numbers of French regulars, but mainly with Indian forces loosely aligned with the French: Haidar Ali was said to have around 70–100,000 infantry in 1780, and about 30,000 cavalry, while the European and sepoy garrison at Pondicherry was less than 1,000 strong in 1778, and in 1783 the marquis de Bussy commanded no more than 2,600 French troops. True, about 30,000 regulars were gathered in Normandy and Brittany in 1778 and 1779 for a projected descent on southern England; but this never occurred. An estimated 20,000 fought in America and the West Indies, and others saw service in Africa, India and Europe (at the sieges of Gibraltar and Minorca); but the majority of France's soldiers remained at home.

In the absence of a major land war in Europe, the French were free to devote their main efforts to a maritime and colonial struggle. Their navy accordingly acquired a new importance. In the past the French had neglected their maritime capacity and concentrated on the army. In 1760 France had spent the equivalent of about £7 million on her land forces, and a mere £500,000 on her navy. But since the end of the Seven Years' War there had been a diversion of money to construct a fleet capable of challenging British seaborne supremacy. Once hostilities commenced, resources were poured into naval expansion: 35 million *livres* (about £1.5 million) were spent on the navy in 1776; 200 million *livres* (nearly £9 million) in 1782. Naval appropriations, moreover, outstripped those for the army: between 1778 and 1782 just under 692 million *livres* were devoted to the navy, as compared with 531.6 millions to the war department. And this extra naval expenditure bore spectacular fruit: France's 52 ships of the line in 1778 had become 73 by 1782.

The French war effort, considered in relation to her population and wealth, was clearly more limited than either the British or the American. Yet considerable strain was imposed on state finances by the American struggle. This was because the French government had no effective means of tapping the resources of the kingdom. Tax exemptions secured by certain provinces and by the noble class throughout France reduced potential tax revenue. The nobility, in defence of their privileges, determinedly resisted fiscal reform. Using their domination of the benches of the Paris and provincial *parlements*, they blocked or mangled the initiatives of the king's ministers. The *parlements*, which were courts rather than representative and legislative bodies, registered royal edicts to give them the full force of law. In return they claimed the right to comment upon, or even to amend, the measures put before them. If the *parlements* proved particularly recalcitrant, edicts could be registered in a special session without their approval; but this was a last resort that ministers tried to avoid. In the final years of the reign of Louis XV a serious attempt was made to remove the obstacles created by the *parlements*: indeed, in 1771 Chancellor Maupeou actually abolished them altogether and introduced new courts. The chance to build on this radical approach was surrendered, however, when the new king came to the throne in 1774. Anxious to begin his reign by healing old wounds, Louis XVI naively restored the *parlements*. Turgot tried again to remodel the tax system, only for the Paris *parlement* to squeal at the prospect of new and more effective imposts. Louis gave way and turned to a less confrontational way of raising money. Jacques Necker, a Swiss with extensive contacts in the banking world, was appointed director of finances in the expectation that he would be able to negotiate loans to pay for a new war with Britain. He was at least partially successful, producing 520 million *livres* (or £21.7 million) between 1777 and 1781. But these sums were raised only by borrowing over relatively short periods at what were in the eighteenth century high rates of interest – up to 10% a year. In January 1781, to reassure the government's creditors, Necker produced the famous *Compte rendu au Roi*. This showed a modest surplus on the king's ordinary accounts; but disingenuously omitted the extraordinary accounts from which most of the interest payments had to be met. Recent scholarship suggests that Necker's wartime borrowing was not as disastrous as historians used to think.[53] Even so, the edifice of French public finance rested on decidedly rickety foundations. A 10% surcharge on most existing taxes had to be introduced in August 1781; and the next year saw a new tax on land to run until three years after the end of the war. That this new imposition was accepted by the *parlements* seems strange until one realizes that it was won at the cost of the government abandoning its attempts to reassess land values – the key to more effective tax collection. The war terminated with the French monarchy no nearer to solving its fiscal problems.

The financial repercussions of the American conflict were to affect France up to the Revolution of 1789, but for most of the war Necker's borrowing papered over the cracks. During the struggle itself, the military effectiveness of the troops and ships brought into action was more important. The reality behind the figures, it could be argued, was less impressive than the figures themselves. France's naval build-up still left her a long way short of parity with the British. France, moreover, had no naval tradition to compare with Britain's; consequently she lacked the quality of dockyard facilities and the same pool of skilled and experienced mariners on which to draw for naval recruits: in 1777 there were probably fewer than 60,000 experienced seamen in France, and nearly 14,000 of them were already in the navy. Manning the expanded fleet caused great difficulties. By 1781–2 there were 90,000 sailors in the French navy; but fewer than half of them were trained French seamen – the rest were foreign mariners, landsmen (conscripts and volunteers) and soldiers drafted from the army.[54]

The army itself continued to exhibit many of the undesirable qualities that had brought it into such disrepute in the later stages of the Seven Years' War. In the previous century Louis XIV had encouraged the nobility to direct their energies to the army, and to accommodate them many more commissions were created than purely military considerations could justify. By 1775 there was one officer for every 15 men, and nearly 1,200 generals (Prussia, with a similar sized army, somehow managed with a miserly 80). This bloated French officer corps was not only expensive; it made much more difficult the attainment of a reasonable level of military efficiency. The higher ranks were full of senior members of the aristocracy who owed their military commands to birth and connections rather than professional merit. Lower down the scale, officers also drawn almost exclusively from the aristocracy tended to be so jealous of their honour and dignity that they were inclined to disregard orders and freely censured their superiors. Nor should we forget that an overly-large officer corps meant an excessive baggage train: noble officers were especially reluctant to forgo their pleasures and comforts, even on campaign. The result – inevitably – was a slow-moving army.

None the less, there is no denying the improvements made since 1763. The French army's poor performance in Germany in the last war led to new thinking in several areas. Jean-Baptiste Gribeauval revamped the artillery; a new infantry musket was introduced which in modified form remained the standard issue for the French army until 1842; the infantry regiments were reformed and standardized; the period of military enlistment was fixed at eight years, with a bounty paid to all who re-joined. Lord Lewisham, a young British aristocrat touring France in 1775, concluded a eulogistic report on this new model French army with the words 'They are really a very noble body of men'.[55] The reformed French navy elicited similar praise. After the clash between the British and French at Ushant, Lieutenant William Home of the Marines wrote from Plymouth: 'It is agreed by every body that no fleet could go thro' the different motions better than what the French did, that they were even superior to the British in that point'.[56]

The French did not always work well with their American allies. At Rhode Island in 1778 and Savannah in 1779 they determinedly pursued their own interests and left the Americans in the lurch. A sense of smouldering resentment was the legacy. When the comte de Rochambeau's expeditionary force arrived at Newport in 1780, the welcome was less than rapturous: 'We did not meet with that reception on landing, which we expected and which we ought to have had', Lieutenant-Colonel comte Guillaume de Deux-Ponts recalled.[57] Complaints were registered against French troops operating in Virginia in 1781; they plundered to such an extent, according to one aggrieved local, that they seemed 'disposed to make the situation of the people more miserable than the British left them'.[58] Few French soldiers or sailors appear to have embraced the American cause; they merely did their duty as they saw it. A junior officer in the artillery could offer only a distinctly simplistic version of events in America: 'In this country one distinguises two parties called Whigs and Tories. The Whigs are called good Americans ... The Tories are known by the name of Royalists';[59] while the chevalier de Chastellux, a *marechal de camp* in Rochambeau's corps, unconsciously revealed his distance from the egalitarian tendencies of the Americans by regretting that circumstances had obliged the British General Earl Cornwallis ('a man of ... birth and character') to take charge of a loyalist 'rabble'.[60]

The British, of course, made much of Franco-American differences. Admiral James Gambier wrote gleefully in September 1778 of how 'The French and rebels are most cordially sick of each other, a most reciprocal enmity and contempt'.[61] But British hopes that the alliance would collapse in a welter of bickering and acrimony were ill-founded. Relations were not as bad as they sometimes appeared. On the march south to Virginia in August 1781, American Brigadier-General Jedediah Huntington noted that 'great Sociability and Harmony prevail between the French Troops and ours'.[62] And so far as ill-conduct towards civilians was concerned, great energy was expended to minimize difficulties. A French grenadier officer whose unit served alongside American troops near New York in July 1781 was shocked by the predatory activities of his allies. 'I had no idea war was waged this way', he wrote. The British and Americans had caught the plundering bug, he concluded, 'but we hope by the force of our discipline to prevent it from happening with us'.[63] In camp at Newport the French troops were effectively isolated from the local population to limit the opportunities for trouble - only the officers seem to have been allowed to move around without restriction. The policy appears to have been successful. Few incidents occurred, and those that did were covered up. From Newport Robert Crook wrote to his sister in the most glowing terms of the French army: 'ther[e] never were troops in Any Country that behaved more Orderly and gave such General Satisfaction'.[64]

*

Spain joined the war not to help the Americans but to regain possessions lost to the British – at the end of the Seven Years' War in the case of the Floridas; and much earlier in the case of Minorca, Gibraltar, Belize, the Mosquito Shore in Central America and Jamaica.

On paper the Spanish army was a significant addition to the allied forces. With 45 regiments of infantry in 1778, 14 of cavalry, eight of dragoons – not to mention guards, artillery and engineers – its full strength was over 70,000 officers and men. It was supported, moreover, by about 23,000 part-time militiamen at home and units of locally-raised troops in the colonies. But in reality the Spanish army was seriously undermanned: through the 1770s and 1780s its true size probably hovered around 30,000, or less than half its official establishment. In the navy, reformed with French help after the Seven Years' War, theoretical strength and real strength more nearly coincided. The tally of actual seamen and marines may have been well down on the 62,000 that appeared in the estimates for 1778, but the number of ships available can be measured with more certainty. In 1776 Spain possessed 57 ships of the line. According to an 'Estado de los Navios', undated but almost certainly from 1779, by the time she entered the American war there was one ship of 120 guns, one of 90, seven of 80, 46 of 70, seven of 64, and two of 60: a total of 64.[65] But Spanish naval strength – in terms of ships of the line actually in service – declined in the course of the war: in 1780 the figure fell to 48, and in 1781 and 1782 it recovered only to 54. So while Britain and France greatly increased their fleets during the conflict, Spain was not even able to mark time.

Significant expansion was ruled out by the limited financial resources available once Spain entered the war. In 1778 the Spanish army, navy and militia received in theory 303 million *reales* (about £8,332,500).[66] Increasing this estimate was no easy matter. Access to the silver and gold of the Spanish American colonies was disrupted by the conflict, forcing the government to rely on domestic revenues. Increases in tax income were difficult to secure. As in France, the root cause of the problem was the inability of the government effectively to tap the kingdom's wealth. One should not exaggerate the similarities to France: the Spanish nobility and clergy lacked concentrated or institutionalized bases of resistance to reforming monarchs and ministers – the *cortes* of Navarre and Castile were much more pliant than the French *parlements*, and the *cortes* of Aragon had ceased to meet. Nor were the privileged classes exempt or immune from taxation. But they escaped lightly. The fiscal regime was sharply regressive, the bulk of state revenue coming from a bewildering variety of excise duties and purchase taxes. The yield was disappointingly low, not least because collection was difficult to police and many of the people eeked out a living beyond the market economy. Numerous attempts were made to rationalize the revenues and to introduce a single tax based on the principle of ability to pay. None was really successful. The reforming zeal of the marques de Esquilache (or marchese di Squillace, to give him his proper Neapolitan name) provoked such an outcry as to force Carlos III to dispense with his services in 1766. Thereafter, all the king's leading ministers – Aranda, Campomanes, Floridablanca – were more cautious. They pushed through some useful modernizing measures, but they never felt confident enough to tackle Spain's fundamental fiscal problem. Taxes on consumption were increased, and new ones introduced – notably on tobacco – but they did little to help. As in France, borrowing was expected to make up the shortfall; and, as in France again, it was often at rates of interest distinctly unfavourable to the state.

Whether the quality of the Spanish armed forces gave value for money is debatable. The Spanish navy, when added to the French, put the Bourbon powers ahead of the British in terms of ships of the line. Yet Spanish naval officers were strongly criticized by their French colleagues for their poor training, low standards of seamanship and inadequate tactical skills. The army's performance was equally open to censure. Years of garrison duty in the same area had led to the rustication of many regiments, and General Alejandro O'Reilly's expedition to Algiers in 1775 was a disaster: badly planned and executed, it ended in humiliating defeat. But perhaps we should not read too much into this unfortunate episode. An English observer with military experience attributed the Algiers fiasco to 'some sad mismanagement', and proclaimed that 'There are no troops in the world more steady than the Spaniards'.[67] The brilliantly successful campaigns of Bernardo de Gálvez in West Florida in 1779–81 – drawing partly on troops and ships from Europe – would seem to give substance to this more optimistic assessment.

*

The Dutch did not so much enter the American war as have the war thrust upon them. It was the British, as we shall see, who forced the United Provinces into the conflict. The result was that the Dutch had no war aims to speak of, other than to hold onto their far-flung commercial posts and colonies and to establish the rights of neutral shippers in wartime.

The anti-British coalition was further strengthened when the Dutch joined it, but the Dutch were a less formidable enemy than the Spanish, let alone the French. The well-established picture of Dutch decay in the eighteenth-century 'Periwig Period' has been criticized by some historians as overly gloomy: the revisionists point to a 'relative' rather than an 'absolute' decline. Even so, there was certainly a dramatic slide in the republic's ability to wage war. The Dutch had made their reputation as a maritime people – as carriers, as fishermen, and as ship-builders – yet in the early 1780s Admiral Joan Arnold Zoutman complained of the enormous efforts needed to find 3,000–4,000 sailors to man the republic's far from sizeable navy. In 1781 the Dutch had a mere 13 ships of the line; in 1782 this had been increased, but only to 19. Nor was the quality impressive: just before the Dutch came into the war their most modern ship of the line dated from 1753.

A good deal of the military effort of the United Provinces in the American war was in fact provided by the forces of the Dutch East India Company, the *Vereenigde Oost–Indische Compagnie (VOC)*. In 1779 the company had about 36 naval vessels in the East Indies, and about 11,000 troops, around 2,700 of whom were based in Ceylon. Further soldiers were hired from local princes, mainly in Indonesia: when the French retook Trincomalee from the British in August 1782, Malay auxiliaries in Dutch pay formed part of the investing army. But although the *VOC*'s strength on paper appeared worthy of respect, and it provided financial backing for French military operations in India, the Dutch company's forces were in reality weak and ineffectual. The *VOC* was able to offer little opposition to British attacks on their trading posts, and had it not been for timely French intervention the base at the Cape of Good Hope would probably have been lost as well. The truth was that the *VOC* was no longer the great power that it had been in the seventeenth century. Since around 1724 it had failed to make an operating profit, and by the time of the American war the company was heavily in debt.

The regular Dutch army was in just as sorry a state as the *VOC*'s forces. It had shrunk to a fraction of its former size. In 1710, during the War of Spanish Succession, the United Provinces had fielded an army of 100,000 men. By the mid-1770s the Dutch land forces had fallen below 30,000 strong. The once-proud Scots brigade, still officered by Scots but now with a cosmopolitan rank and file, could muster no more than 1,800 troops in 1775. As with the navy, reliability was questionable. Military service was unpopular in the republic itself, and to make it more attractive a generous leave policy was adopted: at any given time half the army was likely to be absent on furlough. The consequences were a lack of professional skill and a marked reluctance to be available for active service. In 1778, a year of political tension in Europe, the British ambassador reported that 'The delay in granting furloughs this summer ... has already occasioned a desertion which is computed to amount to nearly a quarter of the army'.[68] Then there was the anamolous situation of the Scots brigade. Its officers technically owed allegiance to the British monarch and served with the Dutch on the assumption that Britain and the United Provinces were allies. Once war broke out between them, the Scots brigade could not be entirely relied upon by the Dutch. It was confined to garrison duties in the so-called Barrier fortresses in the Austrian Netherlands (roughly speaking, modern-day Belgium) until in 1782 it was disbanded, or rather absorbed into the Dutch army proper. New uniforms were issued – Dutch blue coats symbolically replacing the red traditionally worn by the Scots brigade – and the officers were required to swear an oath abjuring allegiance to their native land. Many resigned their commissions and returned to Britain rather than accept these new terms of service.[69]

Pre-war attempts to enhance the military capacity of the republic were stymied by internal political division. The Anglophile *Stadhouder*, Willem V, the Prince of Orange, wanted to augment the army to counter possible French aggression; while his opponents, led by the merchants of Amsterdam, saw the British as the threat and wanted to rebuild the navy to protect Dutch commerce. The Dutch constitution presented a formidable obstacle to Willem. The loose federal arrangements settled by the

Union of Utrecht (1579) meant that the states of each province had to agree to military expenditures and the raising of troops. While the landward provinces were willing to accept Willem's plea for more soldiers, the seaward ones were not. The *Stadhouder's* compromise – a simultaneous increase in both army and navy – was rejected by the powerful province of Holland. Opposition to army expansion was partly ideological and historical – many Dutchmen were temperamentally reluctant to give any Prince of Orange too big a standing army, just in case he tried to emulate Willem II's attempted *coup d'etat* of 1650. Holland's opposition, however, owed much to her merchants' perception of the British as maritime and commercial rivals. And there was also concern at the already-high level of taxation. Resistance to further taxation was particularly fierce in Holland, which contributed nearly 60% of all Dutch revenue. But resistance was by no means confined to Holland. The long wars against the French at the end of the seventeenth century and the beginning of the eighteenth had all but ruined the republic, and they served as a reminder of the folly of a small country trying to maintain military and naval forces beyond her capacity.

*

If we adopt a different perspective, and view the war as a contest between two sides, what can we conclude? While the conflict was simply an Anglo-American struggle, the British were, in theory at least, in a very strong position. The population of the British Isles far outstripped that of the rebel colonies (in 1775 by about 12 millions to 2.5 millions); from 1776 the army under General Howe was larger than Washington's; and the Royal Navy enjoyed a massive superiority over a Continental Navy that lacked any ships of the line. In 1778, however, with the entry into the war of the French, the British advantage started to disappear. France's population was about twice that of Britain and Ireland; France's peacetime army was more than double the size of Britain's war establishment in 1778; and the French fleet, though smaller than the British, was able to mount a real challenge to the Royal Navy. From 1779, when the Spanish joined the fray, even the naval balance tipped in favour of the allies; and in 1781 Dutch involvement made the situation look even more bleak for the British:

Ships of the line of the belligerents 1778–82

	Britain	America	France	Spain	Dutch	Total allied
1778	66	0	52	–	–	52
1779	90	0	63	53	–	121
1780	95	0	69	48	–	117
1781	94	0	70	54	13	137
1782	94	0	73	54	19	146

Source: Jonathan R. Dull, *The French Navy and American Independence* (Princeton, N.J., 1975), pp.359–76.

Yet we should resist the temptation to assume that such a balance of forces made the outcome of the war inevitable. Raw data of this kind can be misleading. Overall numerical advantage does not always translate into local superiority; it can be vitiated by logistical difficulties; and quality, it hardly needs to be said, is often as important as quantity. Morale, like skill and leadership, is not readily reducible to statistical measurement. And we should not forget that fiscal capacity is vital in any war. In this respect Britain retained the upper hand. The conflict imposed great strain on Britain's tax and credit system, but Britain seems to have been better able to sustain a protracted struggle than were France, Spain, the United Provinces, and – arguably – the United States. There is no iron rule, in short, that an outnumbered army or navy is doomed. The Americans successfully held the British at bay even before the French intervened; and after 1778 the British managed to avoid overwhelming defeat despite confronting a formidable coalition that possessed a clear superiority in numbers of fighting men and ships.

Chapter 3 Notes

1. For fuller discussion of the reasons for the slowness of naval mobilization see Nicholas Tracy, *Navies, Deterrence and American Independence* (Vancouver, 1988); and Daniel Baugh, 'The Politics of British Naval Failure, 1775–1777', *American Neptune*, lii (1992), 221–46.
2. George Foley to Sir Richard Heron, 28 Nov., 10 Dec. 1782, Stubton Papers, II D/3, Lincolnshire Archives Office.
3. John Brewer, *The Sinews of Power: War, Money and the English State, 1688–1783* (London, 1989), p. xvii and *passim*.
4. Samuel Kenrick to James Wodrow, 12 May 1780, Wodrow-Kenrick Correspondence, MS 24.157 (69), Dr Williams's Library.

5. Barrington to Gage, 22 June 1775, Letterbook 1766–75, Barrington Papers, HA 174/1026/107, pp. 304–5, East Suffolk RO.
6. Maj.-Gen. Robert Cunninghame, 7 June 1775, WO 27/35.
7. Osborne to [Howe], 23 June 1777, Carleton Papers, PRO 30/55/5, 591.
8. Johann Ewald, *Diary of the American War*, ed. and trans. J.P. Tustin (New Haven, Conn., 1979), p.68.
9. Gray to Capt. Francis Le Maistre, 24 June 1779, WO 28/5, fo. 14.
10. Clinton to Germain, 23 Oct. 1778, *DAR*, xv. 228.
11. *The Letters of Sir William Jones*, ed. Garland Cannon (2 vols., Oxford, 1970), i. 264.
12. Lt.-Col. James Bruce to Lord Ailesbury, 2 Aug. 1778, Savernake MSS, 9, Wiltshire RO.
13. James Falconer to the Earl of Dartmouth, 17 March 1783, Dartmouth MSS, D(W) 1778 v. 731, Staffordshire RO.
14. Coote to Charles Jenkinson, 29 Jan. 1780, Liverpool Papers, Add. MS 38,404, fo. 24.
15. Entry Books Criminal, State Papers Domestic, SP 44/93, p.324, PRO.
16. Lee to Washington, 10 May 1776, 'Lee Papers', New-York Historical Society, *Collections*, v (1872), 18.
17. Henry Strachey to ... [26 Aug. 1776,] (draft) Strachey MSS, DD/SH, C.1165, box 34, Somerset RO.
18. Maj.-Gen. Alexander Leslie to Clinton, 29 May 1781, Clinton Papers, William L. Clements Library.
19. Dundas to James Dundas, 13 May 1780, Dundas of Ochtertyre Muniments, GD 35/57/9, SRO.
20. O'Hara to Sir Charles Thompson, 20 Sept. 1778, Hotham Papers, DD HO 4/19, Hull University Library.
21. 'Extracts from the Letter-Book of Captain Johann Hinrichs', *Pennsylvania Magazine of History and Biography*, xxii (1898), 146.
22. 'The Aspinwall Papers', Massachusetts Historical Society, *Collections*, 4th series, x (1871), 795.
23. 'The Journal of William Humphrey', *Rhode Islanders' Record of the Revolution*, ed. Nathaniel N. Shipton and David Swain (Providence, R.I., 1984), p.41.
24. 'Journal of Du Roi the Elder', trans. Charlotte S.J. Epping, *German American Annals*, xiii (1911), 154.
25. 'Diary of Enos Hitchcock', ed. W.B. Weedon, Rhode Island Historical Society, *Publications*, vii (1899), 169–70.
26. See below, Chapter 9.
27. For more on this theme see below, Chapter 7.
28. George Washington Papers, 6B vol. I, Library of Congress.
29. Seventh Foot Orderly-book, entry of 21 June 1778, Historical Society of Pennsylvania.
30. Maj.-Gen. James Robertson to Lord Amherst, 7 Jan. 1777, Amherst Papers, U 1350 079/14, Kent Archives Office.
31. Howe to John Robinson, 2 Dec. 1776, Treasury Papers, T 64/108, fo. 73, PRO.

32. Clinton to the Duke of Newcastle, 26 Nov. 1778, Newcastle of Clumber MSS, NeC 2646, Nottingham University Library.
33. David Syrett, *Shipping and the American War* (London, 1970)
34. 'George III and the Southern Department: Some Unprinted Royal Correspondence', ed. Ian R. Christie, Royal Historical Society, *Camden Miscellany*, xxx (1990), 431.
35. *Greene Papers*, iv. 108.
36. *Washington Writings*, iii. 307, 450–1.
37. 'The Heath Papers', Massachusetts Historical Society, *Collections*, 7th series, lxiv (1904), 12.
38. 'The Wyllys Papers', Connecticut Historical Society, *Collections*, xxi (1924), 467.
39. Dunglass to the Earl of Home, 19–28 Oct. 1777, Douglas-Home Muniments, box 187, bundle 5, The Hirsel.
40. Ewald, *Diary*, p.108.
41. Percy to Harvey, 20 April 1775, Percy Papers, vol. L, Pt. A, fo. 53, Alnwick Castle.
42. *Papers of John Adams*, ed. Robert J. Taylor (8 vols. to date, Cambridge, Mass, 1977–), v. 21.
43. *Letters and Papers of Major-General John Sullivan*, ed. Otis G. Hammond (3 vols., Concord, N.H., 1930–9), ii. 225.
44. *Hamilton Papers*, ii. 385.
45. 10 Nov. 1780, Letter-book vol. I, Greene Papers, Library of Congress.
46. Greene to Maj.-Gen. James Varnum, 9 Feb. 1779, *Greene Papers*, iii. 223.
47. *The Letters of Lafayette to Washington, 1777–1799*, ed. Louis Gottschalk (Philadelphia, 1976), p.195.
48. Greene to Gen. Robert Howe, 29 Dec. 1780, Letter-book vol. I, Greene Papers; *The Revolutionary Journal of Baron Ludwig von Closen*, trans. and ed. Evelyn M. Acomb (Chapel Hill, N.C., 1958), p.89.
49. Greene to the board of war, 6 Dec. 1780, Letter-book vol. I, Greene Papers.
50. See, for example, David French, *The British Way in Warfare* (London, 1991), pp.71–2.
51. *American Campaigns*, i. 29.
52. *The Papers of George Mason*, ed. Robert A. Rutland (3 vols., Chapel Hill, N.C., 1970), ii. 696.
53. See particularly Robert Harris, *Necker: Reform Statesman of the Ancien Regime* (Berkeley, Ca., 1979), and the consideration of Harris's defence of Necker in Chapter 10 below.
54. For France's difficulties in transferring resources to the navy see Michael Duffy, 'The Foundations of British Naval Power', *The Military Revolution and the State*, ed. Michael Duffy (Exeter, 1980); Daniel A. Baugh, 'Why did Britain lose Command of the Sea during the War for America?', *The British Navy and the Use of Naval Power in the Eighteenth Century*, ed. Jeremy Black and Philip Woodfine (Leicester, 1988); N.A.M. Rodger, 'The Continental Commitment in the Eighteenth Century', *War, Strategy, and International Politics: Essays in Honour of Sir Michael Howard*, ed. Lawrence Freeman, Paul Hayes and Robert O'Neill (Oxford, 1992).

55. HMC, *Manuscripts of the Earl of Dartmouth* (3 vols., London, 1887–96), iii. 219.
56. Home to William Home, 21 Aug. 1778, Logan Home of Edrom Muniments, GD 1/384/6/30, SRO.
57. *My Campaigns in America*, trans. and ed. Samuel Abbott Green (Boston, 1868), p.91.
58. William Reynolds to Gov. Thomas Nelson, 16 Nov. 1781, *Calendar of Virginia State Papers*, ed. William P. Palmer *et al.* (11 vols., Richmond, 1875–93), ii. 601.
59. 'Rhode Island in 1780', trans. E.R. Massey, Rhode Island Historical Society, *Collections*, xvi (1923), 72.
60. *Travels in North America*, trans. and ed. Howard C. Rice Jr. (2 vols., Chapel Hill, N.C., 1963), ii. 570.
61. *Sandwich Papers*, ii. 308.
62. 'Huntington Papers', Connecticut Historical Society, *Collections*, xx (1923), 448.
63. Quoted in Lee Kennett, *The French Forces in America, 1780–1783* (Westport, Conn., and London, 1977), p.120.
64. 15 April 1781, Kingston Collection, box 6, New York State Library.
65. Miscellaneous Papers relating to the Spanish Navy, 1740–82, Add. MS 20,962, fo. 324.
66. Joseph Townsend, *A Journey through Spain in the years 1786 and 1787* (3 vols., London, 1791), ii. 185–6.
67. Philip Thicknesse, *A Year's Journey through France, and Part of Spain* (2 vols., Bath, 1777), i. 164.
68. Quoted in Simon Schama, *Patriots and Liberators: Revolution in the Netherlands 1780–1813* (London, 1992 edn.), p.57.
69. See *Papers Illustrating the History of the Scots Brigade, vol. ii : 1698–1782*, ed. James Ferguson (Edinburgh: for the Scottish History Society, 1899).

4

Military Operations in America, 1775–7

News of Lexington and Concord passed with remarkable speed through Massachusetts, New England, and then the rest of the colonies. Stories of the rapacity and cruelty of the regulars spread like wildfire. On 20 April Abner Sanger of New Hampshire recorded in his journal that his local town was 'in an uproar'.[1] Militiamen from the New England provinces – and eventually from farther afield – converged on Boston to reinforce the local besiegers. Gage's army was cut off from the surrounding countryside, and in a matter of days fresh provisions were running short.

The British position was unenviable topographically as well as logistically. Boston was commanded by high ground, particularly on Charlestown Peninsula to the north and Dorchester Neck to the south. On 19 April, when Percy's battered column reached safety, Admiral Samuel Graves urged the occupation of Charlestown Peninsula and the burning of Charlestown itself. Gage refused, pointing to the limited number of troops at his disposal. He was no more willing to move when Graves offered to land all the marines from the fleet. Gage's reluctance to escalate the conflict no doubt owed much to a wish to keep the way clear for a negotiated settlement. But the arrival from England of three generals – William Howe, Henry Clinton, and John Burgoyne – seems to have persuaded him to take action. After protracted deliberations, a decision was made to seize Dorchester Neck on 18 June. But news of the plan leaked out, and the besiegers took pre-emptive measures. As a first step, Artemas Ward, commander of the investing army until Washington's arrival, determined to fortify Charlestown Peninsula. On the night of 16–17 June, militiamen occupied Breed's Hill and constructed a redoubt.

From this post the Americans could bombard Boston, as they demonstrated by lobbing a few shots into the north end of the town. The British abandoned their preparations for a move against Dorchester Neck, and 2,200 troops under Howe's command were landed on the Charlestown

Peninsula to clear the militia from Breed's Hill. Despatching his light infantry companies to outflank the Americans along the shore of the Mystic River, Howe decided to lead the bulk of his force in a frontal attack. The light infantry, advancing in a column along the narrow beach by the Mystic, were mown down by militiamen waiting behind a hastily constructed stone barricade. Howe's troops, moving forward over difficult terrain, and fired on by snipers in Charlestown, suffered just as badly. Their first assault was driven back, much to the consternation of Howe and his officers. Reinforced by a further 400 troops from Boston, the British attacked again. According to some reports, the American position was taken at the second attempt; others say the third. Either way, 'It was with a Great Difficulty', as one British participant put it, that 'o[u]r men master'd the Field'.[2] They did so only at appalling cost. British casualties amounted to 226 dead and 828 wounded: over 40% of the attacking force. The Americans lost 140 killed and 271 injured. 'Tho our enemies has gain'd ground upon us, it is at too dear a rate for them to rejoice much at their success', wrote Nathanael Greene. 'I wish we could Sell them another Hill at the same Price', he told his brother.[3]

The battle of Bunker Hill, as it was somewhat misleadingly to be called, profoundly shook the British. They had been dramatically disabused of the misapprehension that Americans would be unable to resist advancing regular troops. Continued bragging that the colonists 'never will meet us on open ground',[4] masked a deepening concern about the difficulties confronting the British. The heavy casualties added to the feelings of gloom: 'the Shocking Carnage that day never will be erased out of my mind', wrote Major Francis Bushill Sill of the Sixty-third Foot.[5] Gage, for many months the victim of a whispering campaign, was now openly accused of pro-American leanings. 'Our Commander in Chiefs Conduct the army seem to be much dissatisfied with', Lieutenant William Braco Gordon wrote on 7 July, 'his partiality to the Yenkies is rather too glaring not to meet with censure.'[6] Morale fell further when dysentery broke out, increasing the strain on those troops who remained fit for duty. Possession of Charlestown peninsula brought little advantage: the soldiers defending it were subject to frequent harassment from American sharpshooters. 'Never had the British army so ungenerous an enemy to oppose', Ensign William Carter of the Fortieth Foot declared indignantly. 'What an infamous method of carrying on a war!'[7]

Washington, who reached the American camp on 2 July, had problems of his own. His troops also suffered from a serious outbreak of dysentery. Trying to bring discipline to his command, Washington believed a British attack could take place at any time. He overestimated the size of Gage's force and its willingness to risk another engagement. Bunker Hill put paid to any idea that the British could force their way out of Boston: 'our army is so small that we cannot even afford a Victory if it is attended with any loss of men' was the opinion of Earl Percy.[8] Like many other officers, Percy thought that the army should leave Boston for New York, where both the topography and the disposition of the people would be more favourable to British operations. Gage was already coming to the same conclusion; but only at the end of August did he ask the colonial secretary for permission to move to New York.

By the time permission was received early in October, the unpopular Gage had been replaced as commander-in-chief by Howe. In view of the lateness of the season, Howe decided to delay an evacuation until the spring. A force was sent to destroy the town and shipping of Falmouth, in modern-day Maine: 'This place was always a nest for Smugglers and the most abandoned of the liberty party so that they are not much intitled to pity' was the stern verdict of Richard Reeve, an embittered customs officer.[9] But apart from this excursion, the Boston garrison settled down to a largely uneventful winter.

*

While the main British army was cooped up in Boston, American and British troops clashed in upper New York and Canada. To pre-empt a descent on the rebel colonies from British-held Canada, on 10 May 1775 New England irregulars under Ethan Allen and Benedict Arnold surprised the 50-man garrison of Fort Ticonderoga, a formidable post when in French hands in the Seven Years' War, but by this stage run down and barely defensible. Advancing northwards, the Americans took Crown Point and its tiny garrison of 12 men. These colonial successes were deeply embarrassing for the British. In the 'last war' a subaltern of the Royal Fusiliers wrote home, the capture of these forts 'cost us many a brave and gallant Soldier: and now we [are] depriv'd of them by a Sett of ragamuffins'.[10] Fort St. John, a post on the Richelieu River only about 30 miles from Montreal, was seized a few days after Crown Point, only to be abandoned on the approach of British troops moving south to block the American advance.

The capture of Ticonderoga and Crown Point had been a local, New England initiative (even though Allen demanded the surrender of Ticonderoga in the name of 'the great Jehovah, and the Continental Congress').[11] But in late June Congress itself ordered Major-General Philip Schuyler, commander of the so-called 'Separate army' (separate, that is, from Washington's forces around Boston) to move into Canada. The American rebellion was to be extended throughout the continent, and

The Northern Campaigns, 1775–81

the French Canadians were finally to be given the freedom denied them by the Quebec Act. The attack was to be two-pronged, for while Schuyler prepared to advance north from Crown Point, Arnold was to proceed with volunteers from Washington's army through Maine to Quebec. Schuyler's army, under the effective command of his subordinate Richard Montgomery, a former British officer, commenced operations on 30 August. Fort St. John was laid under siege from 16 September. By 19 October Fort Chambly had fallen; and on 2 November Montgomery's troops finally induced the surrender of Fort St. John and its 500-strong garrison. The defence had been dogged: the British capitulated only when their food supplies were almost exhausted. In retrospect, we can see that the siege probably saved British Canada by using up valuable weeks of campaigning that the Americans could ill afford to lose. At the time, however, it seemed that Montgomery's force was all-conquering. Eleven days after the fall of St. Johns, Montreal was occupied.

Arnold, meanwhile, had set out in mid-September. His troops made a harrowing journey by boat and on foot through the wilderness of Maine. Arnold noted on 12 October that the 'spirit and industry' of his men 'seems to overcome every obstacle'.[12] By the 27th, however, one of his officers was suggesting that only lack of further provisions in the areas through which they had passed convinced the Americans that they should press on rather than turn back. Arnold arrived with 675 men at the St. Lawrence River on 9 November. Captain Henry Dearbon wrote of how 'a Considerable number' had been 'left on the road Sick or woren out with fatigue & hunger'. Those still with Arnold were 'so fatigued, that they were very unfit for action'.[13] Arnold's force was clearly too weak to mount an immediate attack on Quebec; but early in December he was joined by 300 troops under Montgomery.

Although the combined American force was still small, the British position was parlous. General Guy Carleton, governor of Quebec and commander of the British forces, had sent most of his regiments to Boston in the winter of 1774–5. To defend Quebec city he had 1,800 men, but many of them were militia, newly-raised Scots immigrants or seamen rather than regulars. Nevertheless, Carleton's motley army performed creditably when the Americans launched their assault on 31 December – the day before the expiration of the terms of service of Arnold's New Englanders. In dreadful weather, the Americans were repulsed with heavy losses. Montgomery was killed and Arnold injured. Reinforced over the following months, the Americans remained near Quebec preparing for another attempt. But, from the British point of view, the crisis had passed.

*

For most of 1775, the British had been on the defensive, surrounded in Boston and desperately clinging on in Canada. How were they to put down the rebellion? Concentrating troops in Massachusetts had failed to cow the province, or to provide the military strength necessary to crush the insurgents. It was natural, therefore, that during the siege of Boston consideration should be given to alternative strategies. Various officers – both in Britain and America – argued that a naval blockade was the most effective way to deal with the rebellious colonists. The adjutant general, Edward Harvey, maintained that 'attempting to Conquer A[merica] Internally by our L$^{d.}$ Force is as wild an Idea as ever Controverted Com^{n}. Sense', while Captain James Harris of the Fifth Foot wrote of the need to carry on the conflict 'by sea only'.[14] The secretary at war, Lord Barrington, expressed the same view: even before Lexington and Concord he had recommended a purely naval subjugation of the colonies.

Yet the government committed itself to a major land war. In part this was because it seemed the quickest solution to the problem, and therefore the one least likely to provide an opening for the vengeful French – always a major concern of the British government. Besides, there were the American loyalists to consider. Without a military presence to give them support and encouragement, they would be overawed by the revolutionaries and subjected to the 'tyranny' of Congress. With the help of British troops, on the other hand, the loyalists could play a significant role in restoring legitimate authority. So, with a strongly reinforced army, the British government planned to make 1776 the year in which the flames of rebellion were extinguished. But before offensive operations could begin, Howe's troops had to be extricated from Boston.

*

During the winter of 1775–6, Washington was anxious to attack Boston, but his more cautious colleagues vetoed his plans. In November 1775, however, he took a decision that was to precipitate the departure of the British. He ordered the heavy artillery at Fort Ticonderoga to be transported to Boston. Given the time of year, the distance, and the terrain to be crossed, this was no easy task. But by the beginning of February 1776, Fort Ticonderoga's ordnance was at the disposal of Washington's army. Hitherto, the investing forces had lacked adequate artillery for effective siege operations. Now they posed a serious threat to the town and its garrison.

To forestall any bombardment of Boston, Howe on 13 February sent a detachment to destroy all the houses and cover on Dorchester Neck. He would have been better advised to occupy the high ground, for on the night of 4–5 March the Americans entrenched themselves on Dorchester Heights. The colonists had repeated their feat of fortifying Breed's Hill the previous June, and a re-run of the battle of Bunker Hill seemed likely. Howe ordered a force of troops under Percy to prepare to dislodge the Americans. Mercifully, bad weather led to the postponement of what would surely have been another bloody day for the British army. On the following morning, Howe judged the American works to be too strong and abandoned the attack.

His position was now untenable, and preparations to evacuate the troops were brought forward. Many of the wooden buildings in Boston had already been dismantled for fuel in the winter months, but as the British arranged to depart the destruction greatly increased. Howe was anxious to leave nothing that might be useful to Washington. 'Every vessel which they did not carry off', Boston selectman Timothy Newell noted, 'they rendered unfit for use'.[15] Soldiers and sailors added to the losses of the local people by stealing moveable property. 'My store on Long Wharff broke open again', John Rowe wrote in his diary, 'the behaviour of the soldiers is too bad, 'tis almost impossible to believe it'.[16] By 17 March the last of the ships were loaded and they started to sail away. Their destination was not New York, however, but Halifax, Nova Scotia. Shortage of provisions, and the congestion caused by the need to transport loyal civilians as well as the troops, persuaded Howe that the army should go to a post still securely in British hands.

*

From early May, when the ice on the St. Lawrence melted, the way was open for British and German troops under Burgoyne to augment Carleton's forces at Quebec. The British strategy for 1776, agreed during the siege of Boston, was that a northern army, some 10,000 strong, should clear the Americans from Canada and then proceed down into New York via Lake Champlain and on to a junction with Howe's main army operating in the lower Hudson valley.

Once Burgoyne's troops arrived at Quebec, the Americans, heavily outnumbered, began to withdraw. On 8 June 1776, fresh troops sent by Congress to strengthen Arnold's command clashed with the British at Trois Rivières. The Americans suffered a serious defeat. Arnold ordered the evacuation of Montreal, and his troops treated it as severely as Howe's men had dealt with Boston. The French Canadians, who on the whole had so far remained spectators of the war, now had good reason to be hostile to the Americans. On 18 June the British reached Fort St. John. By the 22nd, Captain Henry Gardner, Burgoyne's aide-de-camp, was able to report: 'I dont believe theres a single rebel left in canada except prisoners'.[17]

The British moved on to the shores of Lake Champlain, where Carleton stopped to construct a flotilla for operations on the lake. This delay represented a lost chance for the British. The Americans were given the opportunity not only to elude their British pursuers, but also to strengthen their posts at Crown Point and Ticonderoga. A sign of the boost given to American morale was the return of scouting parties to make guerrilla attacks. On 25 July, a senior British officer was picked off on the road near Fort St. John. Carleton 'expiated on such a cowardly & cruel manner of carrying on the war' and threatened the assassins with execution by hanging.[18]

Eventually, on 11 October, Carleton resumed operations. His flotilla met and defeated Arnold's squadron off Valcour Island. The Americans again retreated, abandoning Crown Point and doing their best to make it uninhabitable. Carleton, however, after receiving a negative reply to his demand that Ticonderoga surrender, decided to go no farther. His reasoning was that the fortress was too strongly defended to start siege operations so late in the year. The Americans were surprised by his timidity; Lieutenant Rufus Wheeler's diary entries indicate their expectation that the British were bound to attack: 'preparing for a battle', he wrote on 15 October; 'waiting with impatience for their advance' (23rd); 'no enemy came' (25th); 'the enemy still came not as was expected' (26th).[19] Worse still, from the British perspective, was Carelton's decision to withdraw from Lake Champlain and winter his troops in Canada. His caution meant that when the British returned to the offensive the next spring, they had to spend valuable time retracing Carleton's footsteps. Small wonder that the descent from Canada in 1777 was entrusted to the more energetic and enterprising Burgoyne.

*

During the siege of Boston, the British government decided to make some effort to support the strong body of loyalists believed to exist in the southern colonies. Colonial governors and other officials sent frequent reports of the favourable disposition of their people. All that was needed, they suggested, was a British military presence and the Revolution would collapse in their colony. North Carolina, where the Scottish settlers in the back country were reputed to be especially loyal, was selected as the starting point. An expedition made up of regular regiments based in the British Isles would sail from Cork to Cape Fear. Once the southern provinces had been pacified, the army would sail north and join Howe in time for the start of the principal campaign in New York.

Clinton, who was given command of this southern operation, left Boston with a small detachment of troops and arrived at Cape Fear on 12 March. There he waited for the arrival of the bulk of his command which had departed from Cork under Charles, Earl Cornwallis in mid-February. Even at this stage Clinton had his doubts. In none of the southern colonies, he wrote, 'can there be found a Sufficient Number of friends to Support themselves when the troops are withdrawn'.[20] A more immediate problem was that he was too late to help the North Carolina loyalists, who had risen prematurely and been crushed at the battle of Moore's Creek Bridge on 27 February. By the time Cornwallis's troops started to arrive on 18 April, there was very little that could be done in North Carolina. Only by advancing deep into the back country could the British have revived the enthusiasm of the vanquished loyalists; and this would have been a most hazardous undertaking without the necessary horses and waggons to carry provisions. On the coast, rather than a friendly welcome, the British experienced the kind of harassment that was to become all too familiar. William Dansey, in command of the light company of the Thirty-third Foot, told his mother that one of his sentries had been injured 'by three Rebels who came upon their Hands and Knees to get a Shot at him as one wou'd at a Duck (such cowardly Scoundrels have we to deal with)'.[21]

After three weeks, the British left for South Carolina. They arrived off Sullivan's Island, which protected Charleston harbour, to find that the revolutionaries had begun to construct a fort to prevent British shipping approaching the town. Rather than landing immediately, however, the British troops were kept on board ship another week, allowing the Americans time to improve their defences. Finally, the British troops were put ashore on Long Island, in the hope that they might take the American batteries from the rear. But Clinton had been misinformed. The passage from Long Island to Sullivan's was much deeper and wider than he had thought. 'This *fordable* creek', Captain Sir James Murray of the Fifty-seventh Foot complained, 'was at no time or place less than 7 or 8 feet deep. So much was the General prepossessed with the idea of this infernal ford, that several days and nights were spent in search of it'.[22] Apparently uncertain what to do, Clinton allowed his men to tarry on Long Island, 'inactive as ever,' Ensign George Vaughan Hart of the Forty-sixth wrote despairingly, 'looking at the Rebbels making Entrenchments opposite to us all the time'.[23]

On 28 June, Commodore Sir Peter Parker, the naval commander, whose relationship with Clinton had already become strained, decided that he could wait no longer. He informed Clinton that he would use the ships at his disposal to silence the batteries in the fort on Sullivan's Island. Conceived out of frustration and executed with little preparation, Parker's attack was a complete failure. The fire from his ships could not subdue the fort; but the fort's guns wreaked considerable damage on the British squadron. Two attempts to land troops on the eastern end of Sullivan's Island were repulsed by the entrenched Americans. Even Charles Lee, commanding the revoluionary forces in the area, was surprised. He was particularly impressed by the conduct of the garrison of the fort and its commander Colonel William Moultrie. 'I had no idea', Lee wrote, 'that so much coolness and intrepidity could be displayed by a collection of raw recruits'.[24] The British, with more and more men succumbing to sicknesses brought on by the sub-tropical climate, embarked north, bloodied and humiliated, to join Howe's main army.

*

The troops who left Boston in March remained in Nova Scotia through April and May. Halifax was not as bad as Boston, but conditions were far from ideal. Some of the regiments were forced to stay on board ship for want of accommodation. Shortage of provisions pushed up prices: the letters of the army's officers are full of complaints at the high cost of living in Halifax. At the end of May supplies arrived from Britain, but there was still no news of the reinforcements Howe expected. In early June, he could wait no longer and decided to start operations without them if necessary. At this stage, with limited forces at his disposal, his 'principal Object', he informed Percy, was 'the getting Possession of the Town of New York'.[25]

The British embarked for New York on 10 June and reached Sandy Hook, south of New York harbour, on the 29th. Washington was waiting for them. He had been at New York since April, attempting to make the city and its environs defensible. Political considerations persuaded him to try to hold the city, just as they compelled Howe to try to take it. But Washington was at a distinct disadvantage. The British fleet, commanded by Howe's brother, Vice-Admiral Richard, Viscount Howe, controllled the navigable waters around New York City. British troops could be landed in great force almost anywhere. Washington was therefore obliged to scatter his forces over a large area in the hope of countering a British disembarkation.

Howe was at first tempted to put his troops ashore on Long Island. He then changed his mind and on 2 July an advanced party of light infantry and grenadiers took Staten Island. The American defenders, seeing the approaching British ships, had already fled. Over the next few days the whole army was put into quarters on the island. For the first time since the war began, the British found themselves in an hospitable environment.

The people were friendly – 'the inhabitants ... shew the greatest Satisfaction in our Arrival'[26] – and, because of a failure on the part of the revolutionaries to remove the island's livestock, '1500 head of Cattle, and a great number of Sheep' were reported to have fallen into the hands of the occupiers.[27]

It was just as well that the British were content with their new post: they remained there through July. Howe was unwilling to attack Washington before the arrival of Clinton's southern army and reinforcements of Guards and Hessian troops sailing from Britain. Only on 1 August did Clinton join Howe on Staten Island. Three days later, the first of the Guards and Hessians began to disembark. The bulk of the reinforcements from Britain arrived on 12 August, and Howe, now enjoying a considerable numerical superiority, finally decided that the time was ripe to put his army into motion.

On 22 August the first contingent of troops landed on Long Island. They advanced to reconnoitre the American defences along the thickly-wooded Heights of Guan, a range of low hills about two miles in advance of the main American entrenchments at Brooklyn. The British advance guard probed the American position at Flatbush, just south of a break in the hills. By 25 August, the force that Howe intended for the attack had been assembled. Three days of skirmishing at Flatbush had revealed that the American defences there were too strong for a direct assault. Some other means of approaching the lines at Brooklyn had to be devised. Clinton noticed that Jamaica Pass, almost four miles from Flatbush on the far left of the American position on the Heights of Guan, was undefended. An enveloping flanking manoeuvre was planned. The majority of the British troops, led by Clinton, marched via the backroads to Jamaica Pass during the night of 26–7 August. They rested there until daybreak, when they proceeded along the road to Bedford. Two brigades of British troops under Major-General James Grant were meanwhile creating a diversion on the far right of the American position, and a Hessian force under Major-General Leopold von Heister started to shell the defenders of Flatbush Pass. Clinton's force arrived at Bedford, well behind the American centre, at 9 o'clock on the morning of the 27th. They announced their presence by heavy firing. This was the signal for the other British and Hessian troops to attack in earnest. The Americans were taken completely by surprise. Lieutenant Jabez Fitch, on the American right opposite Grant, wrote of how 'we heard a Smart Fire Toward our Left Wing, which gave us to Appprehend that we were in Danger of having our Retreat cut off'.[28] Their centre, commanded by Major-General John Sullivan, folded rapidly. Sullivan himself was captured. The American right, under William Alexander, who claimed the title Lord Stirling, put up more stubborn resistance, and was overcome only when detachments from Clinton's flanking column joined in the attack. The Americans, as Clinton put it, were then 'reduced to a *sauve qui peut*; over bays, marshes, creeks &c'.[29] Stirling joined Sullivan among the prisoners. By midday, the British were poised outside the defences at Brooklyn.

Rather than take the American lines by storm, Howe decided to avoid the risk of heavy British casualties by opening regular siege works. Washington's instinct was to retreat no farther; he even called for reinforcements from Manhatten. On 29 August, however, he was persuaded by his fellow officers that the lines were impossible to hold. If Admiral Howe used his naval superiority to cut off the American forces from Manhatten, Washington's army would be forced to capitulate. On the night of 29–30 August, the Americans were evacuated in a host of small boats – an American anticipation of Dunkirk. The British became aware of the departure of the revolutionary troops only in the early hours of the morning, and 'came up time enough to fire at some of their Rear Guard in their Boats'.[30]

The British had achieved a notable victory – a victory undiminished, in many minds, by Washington's escape. Lieutenant-Colonel John Campbell of the Twenty-second Foot merely remarked that the Americans' talent for constructing defensive works was not matched by a determination to hold the ground they had fortified: 'tho' They work'd like Moles To get under Ground, they ran like Rabbits To get from them'.[31] But some of Howe's officers felt that an opportunity had been missed. Captain William Bamford of the Thirty-fifth Foot was one. Had the American lines been taken by storm on 27 August, he wrote, 'we might easily have put an End to the Rebellion'. Dismayed by Howe's caution, Bamford implied that the war was deliberately being prolonged to enable high-ranking officers in the commissariat and other supply services to make their fortunes.[32] Such allegations were to be heard more and more frequently as the conflict dragged on.

Howe resumed offensive operations on 15 September. After a naval bombardment to soften up the American defences, British troops landed at Kipp's Bay on the east side of Manhatten Island. Again, a chance to cut off Washington's troops was lost. Rather than press on to the Hudson shore, the British concentrated on securing the landing place. The American forces based in New York itself were able to escape up the western side of the island. The British moved south and occupied the city. The next day, 16 September, the British received a minor check. Overconfident light infantrymen were repulsed by an American regiment on the Harlem Heights. The morale of the revolutionaries was accordingly boosted. 'Our troops had the honour of behaving well,' wrote Connecticut officer Benjamin Trumbull, 'and the issue of the Battle gave Spirit to them.'[33] After this engagement, the British became more cautious. Both sides dug in. Washington strengthened his defences on the northern end of Manhatten Island, and the British started to construct parallel works.

Howe decided to use the mobility given to him by his brother's fleet to break the deadlock. He set out again to outflank Washington. British troops were landed on 12 October at Thog's (or Frog's) Neck on the mainland. Washington, recognizing the threat posed to his position on Manhatten, began to withdraw most of his forces from Harlem Heights. The British were unable to move off Thog's neck, as the Americans had destroyed the only bridge linking it to the surrounding countryside. So, a few days later, they re-embarked and landed farther east at Pell's Point. From there they advanced at a leisurely pace to White Plains, where Washington's army was encamped. By the time the British arrived, the Americans were well entrenched. On 28 October, British and Hessian troops attacked Chatterton's Hill on the far right of the American lines. A fierce battle ensued – Rhode Islander Sergeant John Smith recorded that 'the firin on Boath Sides was Verey Warm'[34] – and the hill was taken. But no assault on the rest of the defences was ordered. The disgruntled Captain Bamford bemoaned another missed opportunity.

As Washington withdrew northwards, Howe turned his attention to the revolutionary troops remaining on Manhatten Island. While Howe and Washington had been manoeuvering on the mainland, on 27 October Percy had tested the American lines from the south, only to judge them too formidable for a frontal assault. But now, with Howe's main army to the north and east of the island, and the Royal Navy in command of the Hudson to the west, the Americans were trapped. On 15 November, Howe demanded the surrender of Fort Washington, the central point of the American defences. Colonel Robert Magaw, the fort's commandant, defiantly refused. The next day, the British launched simultaneous attacks from the south, east, and north. The outer defences collapsed, and the Americans, crowded in the fort, surrendered. Their losses, killed, wounded, and prisoners, amounted to more than 3,000 officers and men – which put their defeat on a par with the battle of Long Island.

Howe decided to detach part of his command to take Newport, Rhode Island, and on 1 December Clinton and Percy left New York with 7,000 troops. They took Newport without opposition on 8 December, and in the process blocked most of the Continental Navy in Providence harbour. But the main theatre of operations was now New Jersey. Washington had already taken about 2,000 troops across the Hudson at Peekskill on 9 and 10 November, leaving nearly 9,000 troops under Generals Charles Lee and William Heath to parry any possible British thrust into southern New

England. On 20 November, Cornwallis and 4,000 British troops landed at Closter, New Jersey, and moving quickly south they nearly cornered the garrison of Fort Lee. The British were in high spirits: 'we are able almost to do anything', one officer wrote exultantly.[35] American morale, by contrast, was sinking. From Peekskill, Captain Ebenezer Huntington lamented that 'the present appearance is very Gloomy, the British troops making head wherever they attempt'.[36] On 28 November, Cornwallis's advance guard reached Newark just as the last of Washington's men were leaving. Cornwallis almost caught Washington again at New Brunswick on 1 December. By this time, the retreating American army was melting away. Militiamen from New Jersey and Maryland, whose enlistments expired on 30 November, started to go home. Washington pleaded with the governor of New Jersey to 'fall upon proper means to draw forth the Strength of your province to my support'.[37]

Howe, who joined Cornwallis at New Brunswick, came close to catching up with American stragglers at Princeton on 7 December. But when the British reached Trenton the next day, they discovered that Washington had crossed the Delaware into Pennsylvania. Howe may have considered extending the pursuit and pressing on to Philadelphia. In Pennsylvania, no one had much doubt that this was his intention: 'there is grate expectations of the City of Philad^{a} being attacked by Gener^{l} How^{s} Armey', an alarmed merchant wrote on 6 December.[38] But Howe could find no boats, and on 14 December he ordered his men into winter quarters. There was some talk of continuing the advance once the Delaware froze over, but for the most part the British were content to rest on their laurels after what one officer described as 'a glorious campaign'.[39] Six months of almost uninterrupted success had been crowned on 13 December by the capture of Charles Lee, who had led his troops from New York into northern New Jersey to shadow the British advance. Lee's captor, Lieutenant-Colonel the Hon. William Harcourt of the Sixteenth Light Dragoons, expressing what was surely a general view, looked forward to 'a quiet winter' and the 'speedy conclusion' of the war.[40]

Neither expectation was to be fulfilled. Small parties of revolutionary soldiers remained active behind the British lines. Nor were they the only enemy with which the occupying forces had to contend. It seems that some of the local people had been stirred to action by the treatment they received from the British during their advance across New Jersey – the plundering and rapes provoked extensive censure: 'the devastation they make, wherever they come, is not to be equalled in history', a newspaper commented;[41] 'They to the Disgrace of a Civilized Nation', wrote an outraged American officer, 'Ravish the fair Sex, from the Age of ten to

Seventy'.[42] In revenge, inhabitants started to shoot at isolated groups of soldiers or individual redcoats. Attacks on British troops travelling between posts became so much of a problem that, as early as 12 December, Howe promised to execute 'Without Tryal' any Americans 'not drest like sold^s and without officers' who fired on the British troops.[43] This threat seems to have achieved little improvement in the situation: he was obliged to repeat the order on 23 December, and on that day Ensign Alexander Maitland of the Forty-ninth Foot was still complaining of 'a parcel of rascals ... lurking about in woods & bushes, who fire at or rather murder every body who is unlucky enough to get within their reach'.[44] But worse was to come. On the night of 25–6 December, Washington boldly counter-attacked.

*

Washington selected as his main target the Hessian brigade under Colonel Johann Gottlieb Rall at Trenton. The Hessian post at Mount Holly, commanded by Colonel Karl von Donop, was also to be attacked to prevent reinforcements coming to Rall's aid. The timing could not have been better. In all the Hessian posts, the soldiers were enjoying themselves as best they could.

The raid on Rall's brigade was a great success. Striking early in the morning, the Americans completely surprised the Hessians, many of whom were still recovering from the previous night's celebrations. Elisha Bostwick, a sergeant in the Seventh Connecticut Regiment, recalled that once the Hessian artillery was taken 'they resign'd themselves with little opposition'.[45] More than 100 German troops were killed or injured, and nearly 1,000 captured. Rall himself was among the dead. American losses amounted to only four wounded. The American victory might have been yet more dramatic had the flight of about 500 of Rall's men over Assunpink creek, to the south of Trenton, been cut off as planned. But the troops allocated for this task, and those directed to attack Mount Holly, were unable to cross the Delaware. Washington, fearing a visit from Donop, recrossed into Pennsylvania that same afternoon. He need not have worried. Rall's fate encouraged the Hessians to evacuate all their posts along the Delaware. Re-crossing to Trenton on 27 December, Washington concentrated his forces there for an advance deep into New Jersey. By the end of the year he had around 5,000 men under his command.

Howe, on hearing of Washington's renewed presence at Trenton, ordered Cornwallis to dislodge the Americans. Cornwallis arrived at Trenton late in the afternoon of 2 January 1777, to find Washington's army drawn up along the south-eastern bank of Assunpink creek. With darkness falling, the British commander decided to postpone his attack until the next morning. Washington eluded him in another bold move. Rather than trying to return across the Delaware, he made a night march towards Princeton, leaving a small force along Assunpink Creek to give the impression that his army was still in place waiting to do battle with Cornwallis. Approaching Princeton the next morning, the van of Washington's force met Lieutenant-Colonel Charles Mawhood and two British regiments marching south to join Cornwallis. Mawhood's own regiment, the Seventeenth, attacked with the bayonet. The odds against the British were very heavy, however, and growing heavier as more and more of Washington's force arrived. Mawhood's second regiment, the Fifty-fifth, retreated back to Princeton, where the Fortieth Foot remained ensconsed, refusing Mawhood's request for assistance. The Fortieth and Fifty-fifth attempted to escape to New Brunswick, but many members of these units fell into American hands. Mawhood, with the severely battered Seventeenth, managed miraculously to shake off the Americans and reached Cornwallis at Trenton. In total, the British lost 200 captured and 74 killed and wounded, the vast majority of the dead and injured coming from the Seventeenth Foot.

The American victories at Trenton and Princeton caused the British considerable difficulties, and may have deprived them of their best opportunity to win the war. After Princeton, Cornwallis retreated to protect his main supply depot at New Brunswick. The Waldeckers posted at Elizabeth Town were driven out on 6 January, and on the same day Hackensack fell to American troops. The British were now confined to New Brunswick, Perth Amboy and Piscataway. There they spent 'a very disagreeable Winter, continually harrass'd', as Lieutenant Henry Stirke of the Tenth Foot put it; 'almost daily molested', in the words of Captain William Sutherland of the Fifty-fifth.[46] The pushing back of the British posts to the line of the Raritan River meant that Howe was denied access to many of the supplies he had hoped to draw from New Jersey. This led him to fall out with Earl Percy about the quantity of forage that could legitimately be taken from Rhode Island without distressing the inhabitants. More seriously, Howe's posts along the Raritan were much farther away from Philadelphia than the British had been before Washington's offensive. In the spring, an attack on the city would require another campaign in New Jersey to regain the lost ground. Washington's success also made the Americans a more formidable enemy. Prior to Rall's defeat, Lieutenant Robert Grant of the Seventy-first Foot reflected later, 'they were intirely dispirited' and the rebellion on the verge of collapse.[47] Perhaps British officers, keen to view the Revolution as a narrow conspiracy, were inclined to exaggerate the precariousness of the American position in the days before Washington's counter-attack. After all, Pennsylvanians had rallied to his support before Trenton. But on 18 December Washington himself had confessed that without the new army Congress had authorized, *'the game is pretty near up'*.[48]

*

Burgoyne returned to England during the winter of 1776–7 to lobby for the command of a renewed push from Canada. At first he thought in terms of the northern army taking Ticonderoga and then moving on either south to the Hudson or east into New England, where it could co-operate with troops advancing from Rhode Island. The assault on New England was ruled out by Howe's decision, after Trenton and Princeton, to reduce the force at Rhode Island to increase his army in New Jersey. But the idea of Burgoyne's marching down the Hudson to Albany attracted the king and the American secretary, Lord George Germain. The rebellion would surely collapse if seemingly irreconcilable New England could be cut off from the other colonies. Longer consideration might have cast doubt on Burgoyne's proposal. How would the British maintain an effective blockade along the Hudson, and to what purpose? Were they intending to subdue New England militarily, or to blockade it into submission? But if the end was unclear, a more immediate problem was the means. For Burgoyne to succeed, Howe's army had to offer some sort of co-operation; yet no effort was made to ensure that such co-operation was forthcoming.

Howe appears to have accepted the logic of an expedition from Canada, and his early plans for 1777 were compatible with Burgoyne's grand design. On 30 November 1776, Howe envisaged at least 10,000 troops advancing north up the Hudson to Albany, with only 8,000 men, forming a defensive force, covering New Jersey and keeping Washington in check by seeming to threaten Philadelphia. Less than a month later, he had changed his mind. British success in New Jersey convinced him that his main effort should be directed towards Washington and Philadelphia. In a new plan submitted to Germain on 20 December, Howe allocated only 3,000 men to the lower Hudson; and these troops, besides facilitating 'in some degree the approach of the army from Canada', were also meant to cover operations in New Jersey.[49]

This new plan depended upon substantial reinforcements. After Trenton and Princeton, Howe asked for yet more. Washington's counter-attack had not deterred him from wishing to carry out offensive operations in New Jersey and Pennsylvania: on the contrary, it seemed to intensify his desire to defeat the main American army or, failing that, to take Philadelphia. As Howe's fixation with Washington and Philadelphia developed, Burgoyne's requirements were relegated to the back of his mind. The failure of the government to produce adequate reinforcements might have been expected to cause Howe to reconsider operations in the middle colonies; instead it led him to downgrade still further his commitment to the northern army. On 2 April he informed Carleton that the force due to advance from Canada could expect little help from him, as he would probably be occupied in Pennsylvania when Burgoyne moved south from Ticonderoga. On the same day, Howe told Germain that he planned to approach Philadelphia by sea, rather than force his way across New Jersey. Howe's decision made early co-operation with Burgoyne impossible, and on 20 April Howe confirmed this by casting doubt on his ability to complete the Pennsylvania campaign in time to help in the Hudson valley.

Germain, whose responsibility it was to oversee strategy, should no doubt have instructed Howe to reconcile his plans with Burgoyne's. But Germain was reluctant to override the decisions of military commanders on the spot. As a result, proper co-operation was lacking. What was originally viewed from London as a co-ordinated operation to bring the war to an end, became two separate campaigns, pursued with only vague and intermittent reference to one another.

*

Washington had wintered his army at Morristown, New Jersey, a post from which he could move against a British force moving towards Philadelphia or up the Hudson. Howe's plans remained unknown, even to his senior officers. In April, Colonel Donop told his friend Friedrich Wilhelm, Prince of Prussia, that he expected soon to be moving north to join Burgoyne in an attack on the people of New England, 'the authors of the rebellion'.[50] A raid on Danbury, Connecticut, in the same month destroyed large quantities of munitions that might have been used by the New England troops who would oppose Burgoyne, and was interpreted as a sign of Howe's intention to facilitate the operations of the northern army.

But Howe's attention was focused elsewhere. He wanted above all to bring Washington to battle. On 17 June he moved troops from Staten Island into New Jersey and advanced towards Philadelphia. Washington refused to be drawn, despite the plundering and destruction perpetrated by the British. A number of small-scale actions took place, however, in which local people played a prominent role. In contrast to their behaviour during the first British advance of late November and early December

1776, the militia of New Jersey turned out to support the Continental army. Washington praised 'the most Spirited manner' in which they had rapidly assembled to give the British 'every annoyance in their power and to afford us every possible aid'.[51] Howe withdrew, but returned the next week. This time he almost lured Washington into his trap. Appearing to retreat a second time, Howe tempted Washington to follow with part of his army. Howe then turned on the Americans and attacked. 'The Rebels made no stand, but a running fight of it untill they got to their strong holds', a British participant observed disappointedly.[52] Washington's position at Middlebrook was considered too strong to assault, and so Howe returned to Amboy. From there his army embarked for Staten Island, leaving the whole of New Jersey in the hands of the revolutionaries.

On 9 July, a large force of British troops started to board the ships in New York harbour. The Americans were puzzled about their destination. Some of Howe's officers also felt uneasy. Major Charles Stuart of the Forty-third Foot, not knowing his commander's plans but suspecting that Philadelphia was the objective, trembled for Burgoyne. It seemed that the Canadian army was expected to proceed through 'one of the strongest countries upon this continent ... without any assistance from us'. Stuart hoped that 'what now appears to us a determined project, is but a feint, and that we shall turn our arms to the northward'.[53]

But it was not a feint. On 23 July the fleet finally departed for Philadelphia. Adverse weather made progress very slow. The start of the Pennsylvania campaign was delayed further by Howe's decision not to take the most direct route up the Delaware River. Misinformed that Washington was marching to Wilmington, and would therefore be able to oppose a landing on the lower Delaware, Howe believed it safer to proceed up Chesapeake Bay. His army duly landed near Head of Elk, in Maryland, on 25 August. 'We have met with no enemy as yet, nor any friends', a Royal Artillery officer wrote two days later, 'The country is quite deserted.'[54] The British soon discovered that this was not true: on the 31st Major-General Grant referred to the irritation of 'the Militia in Arms to pop at Straglers'.[55] Even so, there was no substantial American force to oppose them. Only when he heard of Howe's landing did Washington march south through Philadelphia to block the British advance.

Washington hoped to halt the British at Brandywine Creek. He had selected a strong defensive position, for the creek cut through wooded slopes and could only be crossed at easily defensible fords. The centre of the American line was at Chadd's Ford, on the main road to Philadelphia. But far to the right of his position, Washington had left other fords unmanned. Howe, briefed on local geography by loyalists from the neighbourhood, saw the chance to repeat the wide flanking manoeuvre that had paid such handsome dividends on Long Island. Sending a column of troops under the Hessian General Knyphausen to stage a diversion at Chadd's Ford, Howe himself set out with Cornwallis and another strong column at 4 o'clock on the morning of 11 September to cross the Brandywine upstream and roll up the American right. Knyphausen's artillery opened fire at 10 o'clock, and a fierce fight developed as the British sought to expell American advanced parties from the west bank of the Brandywine. By this time, Washington had received warnings of the march of the column under Howe and Cornwallis. Unaccountably, he chose to ignore them. Howe moved slowly - one source says that he was mislead about the distance involved and therefore set out too late. It was early afternoon before his men were across the Brandywine at Jeffries Ford and approaching the American right. Even now, Howe seemed in no hurry. A circuitous 18-mile march had tired his troops. His tardiness gave the Americans the opportunity to redeploy to face the new challenge. They did so without any of the panic seen on Long Island, and they were more or less ready when, at 4 o'clock, Howe finally brought his force into action.

At the same time, Knyphausen's troops, hearing the distant firing, and having already pushed the American advanced parties back across the river, began their attack. Wading through the waist-deep water, Knyphausen's command suffered badly from the determined opposition on the east bank. Undeterred, they advanced steadily and the Americans, with equal steadiness, withdrew. On the American right, meanwhile, Howe and Cornwallis had broken through a gap in the new American line. A confused fight ensued, in which the British at length gained the upper hand. The revolutionary troops gave way and retreated. At the cost of around 550 casualties, the British had killed, wounded, or captured over 1,000 American troops.

Yet the British were in no position to capitilize on their success. The main part of the battle had started too late in the day. 'Had not the darkness favored their retreat,' Major Carl Baurmeister wrote, 'we might have come into possession of much artillery, munitions, and horses'.[56] As it was, the Americans were beaten but not humiliated. Captain Joseph Bloomfield of the Third New Jersey Regiment considered it 'an unfortunate day to our Arms', but far from a disaster.[57] The decisive victory that Howe had sought for so long still eluded him. His army remained on the battlefield. There was no pursuit of Washington's retreating troops. This might partly be explained by the lack of sufficient cavalry at Howe's disposal. More important, perhaps, was the exhaustion of the British troops who had endured a long and hard contest. But Captain Ewald of the Hessian *Jägers,* a member of Howe's flanking column, was inclined to suspect political motives. A few days after the battle he complained: 'I firmly believe that we still could have caught up with the greatest part of the enemy army ... if it had been the will of General Howe'.[58]

Over the next two weeks the British and Americans were involved in a number of small-scale engagements. A clash at Warren Tavern, west of Philadelphia, might have developed into a major action had heavy rain not intervened. Early on the morning of 21 September, a British force under Major-General Grey surprised sleeping American troops at Paoli, near Warren Tavern. The British used their bayonets to kill or injure more than 300 of the Americans, and took another 100 prisoner. Five days later, Howe, having out-manoeuvred Washington, marched into an undefended Philadelphia.

Howe's consolation prize was of a very inferior quality. Philadelphia's capture in 1777 brought none of the advantages expected in 1776. American morale might have been dented, but it was far from fatally undermined. News of successes against Burgoyne had reached Pennsylvania, and did much to counteract the blow. Moreover, Washington's army – increasingly seen as the symbol and guarantor of the Revolution – was still intact and, as it was soon to show, capable of striking back.

Howe camped the bulk of his army out of Philadlephia at Germantown, five miles to the north. On the morning of 4 October, Washington, perhaps seeking to repeat his triumph at Trenton, launched an unexpected attack. His plan called for a co-ordinated advance by four converging columns. It would have been difficult to execute in ideal conditions, but that morning a thick fog shrouded Germantown, reducing visibility to 50 yards. Nevertheless, the attack went well at first. The British piquets retreated and the Americans made good progress. 'The Attack was made with such Briskness that we had the pleasure to pursue the Enemy Intirely through German Town', Captain William Beatty of the Maryland line recalled.[59] The Americans were delayed, however, by the staunch defence of Chew House, a stone mansion occupied hurriedly by Lieutenant-Colonel Thomas Musgrove and around 100 men of the Fortieth Foot. The Americans wasted valuable time trying to take Chew House. They failed to do so, despite using artillery against it, and when they resumed their advance, Howe's main force had rallied and was ready to repel them. The fog, which was becoming thicker, now caused great confusion. American troops in the different columns fired on each other. At this moment, Howe counter-attacked. The Americans, so close to victory early in the day, fell back. 'It was disagreeable to have to leave the field, when we had almost made a conquest', a Pennsylvanian officer ruefully noted.[60] Yet this was no rout. Some of Washington's troops had taken no part in the action, and even those who had been heavily involved were for the most part able to extricate themselves. The losses on both sides were comparable to those on the Brandywine.

Germantown demonstrated Washington's resilience and, while technically a British victory, brought little advantage to Howe. Philadelphia, to which the British now withdrew, was largely denied supplies from the surrounding countryside by the vigilence of Washington's patrols. Worse still, the Americans continued to block the Delaware south of Philadelphia, preventing British ships bringing in provisions. British attempts to shift the Americans from their river defences proved fruitless. Fort Mifflin on Mud Island was bombarded on 11 October, without inducing the garrison to capitulate. On 22 October, a Hessian force under Donop attacked Fort Mercer on the eastern shore of the Delaware at Red Bank. Donop's troops arrived outside the fort undetected; but instead of launching an immediate assault, Donop summoned the Americans to surrender. This offer, while no doubt intended to remove the need for loss of life, proved an expensive mistake. The garrison was alerted and given the opportunity to prepare for the attack. As British naval vessels were unable to break through American obstructions across the river, the fort's defenders could be supported by the cannons on a number of revolutionary gallies. Donop's assault was an heroic but costly fiasco. 'His men advanced with great bravery & firmness', a chaplain in the American garrison confessed; but after forty minutes of deadly fire from the fort and the gallies, 'they were obliged to retreat in great Disorder'.[61] Nearly 400 Hessians were cut down – including Donop himself, who was mortally wounded. The following day, as if to underline British discomfort, two warships were set ablaze while trying to bombard Fort Mifflin. Only in mid-November did the Americans evacuate their defences. By then any joy at the easing of Philadelphia's supply difficulties was unable to cancel out the deep depression caused by the fate of Burgoyne's northern army.

*

In June 1777, while Howe's army was still in New Jersey, Burgoyne's troops started to advance south from St. Johns. His force comprised around 7,400 regulars – British and German; 650 loyalists and Canadians; and about 400 Iroquois Indians. To support his drive to Albany, another force under Lieutenant-Colonel Barry St. Leger was to move down the Mohawk valley from Oswego on Lake Ontario. To carry out this colateral movement, St. Leger was given a similarly mixed command: 1,600 British, provincials, and Indians. The assumption of the British planners was that local loyalists would flock to St. Leger's standard.

Burgoyne's army reached Ticonderoga on 1 July. The Americans had been given ample time to put their defences in good order; but the commandant, Major-General Arthur St. Clair, neglected to occupy Sugar Loaf Hill, a 750-foot eminence that overlooked the fort. St. Clair believed it impossible to pull artillery up the heavily-wooded and steep slopes. By 5 July, however, the British had emplaced cannon on the hill. St. Clair thereupon decided to abandon Ticonderoga, and slipped away that night with his 2,300 strong garrison.

The evacuation of Ticonderoga shocked Congress and increased the undercurrent of criticism directed at the unpopular American commander of the Northern Department, Philip Schuyler. The British were also surprised by the ease with which they took such a stronghold. Burgoyne had brought a considerable artillery train – 138 field pieces and howitzers – in expectation of a long siege. Julius Friedrich Wasmus, a surgeon attached to the Brunswick Dragoon Regiment Prinz Ludwig, was amazed by the flight of the Americans. 'Who could have thought or imagined that the enemy would abandon such an advantageous position!', he exclaimed on 6 July.[62] But St. Clair's evacuation, though criticized at the time, may well have played a part in the final humbling of the British. It made Burgoyne's army over-confident: 'From their quitting that Post before our attack began', Lieutenant Lord Francis Napier of the Thirty-first Foot wrote later, 'We had conceived the Idea of our being irresistible'.[63] Just as important, St. Clair's retreat preserved intact the garrison. Had the British cut him off, and the 2,300 Americans surrendered after a siege, the loss of men would surely have been a still severer blow than the loss of the fort.

The British advanced guard under Brigadier-General Simon Fraser set off in pursuit of St. Clair as soon as the British realized that the Americans had departed. Fraser's corps caught the rear guard of St. Clair's troops at Hubbardton on 7 July. The resistance offered by the Americans showed that their morale had not been seriously dented by the evacuation of Ticonderoga. For a time, indeed, it seemed that Fraser's men might be overwhelmed. Only the arrival of German reinforcements under Baron von Riedesel tipped the scale in Fraser's favour; one British officer noted that members of the advanced corps regarded Riedesel's 'fortunate arrival as a matter absolutely necessary'.[64]

The pursuit of St. Clair had taken Burgoyne away from his preferred route south. He had intended to sail from Ticonderoga down Lake George to Fort George. From there he could take the road to Fort Edward on the Hudson. Instead, he now found himself further east, with his main force sailing down Lake Champlain to Skenesboro, and his advance guard still chasing St. Clair overland. Once the British columns converged at Skenesboro, Burgoyne was faced with the prospect of a tough march, without a proper road, to reach Fort Edward and the Hudson. He might have returned up Lake Champlain and then sailed down Lake George; but, characteristically, he dismissed this as a retrograde step that would undermine the high spirits of his troops.

The Americans continued to retreat. A rumour circulated that Schuyler had deliberately ordered the abandoning of Ticonderoga to help Burgoyne. This was untrue; but Schuyler's pessimism was undeniable. After establishing a post at Saratoga in late July he argued that without substantial reinforcements he had no 'great prospect of holding the ground' he had just occupied.[65] Always willing to stress his own difficulties, he seemed to ignore the problems facing his opponents. In reality, these were as serious as Schuyler's, if not more so. It took Burgoyne's army until 30 July to reach Fort Edward. His troops had been obliged to build around 40 bridges and to clear obstructions put in their way by the retreating Americans. The three week's march had not only exhausted Burgoyne's troops; it had seriously depleted their supplies. Burgoyne's logistical difficulties were so acute that they compelled him to turn his attention from the push south. Under pressure from Riedesel, who wanted to find horses for the dismounted Dragoon Regiment Prinz Ludwig, Burgoyne agreed to send a sizeable force east into the Connecticut valley. On 11 August, Lieutenant-Colonel Friedrich Baum was despatched with 600 Germans, loyalists, and Indians to collect horses and cattle and seek out local 'friends to government'. Baum was confronted near Bennington by a much larger force of New Hampshire militia under Brigadier-General John Stark. The Indians fled, leaving Baum still more heavily outnumbered. His Brunswick troops fought bravely but were forced to surrender on 16 August. A relief force under Lieutenant-Colonel Heinrich von Breymann, sent out by an anxious Burgoyne on the 14th, was also destroyed by Stark's militia.

Bennington was a disaster for Burgoyne's expedition. He had lost 900 valuable troops killed or captured, a reduction in his strength that he could ill afford. And Baum's defeat had an immediate impact on local allegiances. Lieutenant William Digby of the Fifty-third Foot had noted on the arrival of the British at Fort Edward how 'many of the Country people came to us for protection'. On 19 August he wrote that many of the same inhabitants 'went from us over to the Enemy'.[66] Burgoyne's position was in fact even worse than he then realized. St. Leger's supporting force had started to turn back to Canada on 22 August. St. Leger had landed at Oswego on 25 July, and a week later he laid siege to Fort Stanwix. The garrison, under Colonel Peter Ganesvoort, was much stronger than the British had anticipated. The American defence allowed time for the local militia to assemble, and a force led by Brigadier-General Nicholas

Herkimer set out to relieve Fort Stanwix. Herkimer's militia were halted in a fiercely-fought battle at Oriskany on 6 August; but St. Leger's Indians were unhappy at the progress of the siege and made yet more uneasy by reports of the approach of a body of Continentals under Benedict Arnold. When the Indians left St. Leger's camp, he was forced to lift the siege and return to Oswego.

As the British position deteriorated, American prospects improved. Horatio Gates took over from the gloomy Schuyler on 19 August, by which time the steady inflow of militia from New England and Continentals from the south had increased his army to around 7,000 men; rather more, in other words, than Burgoyne's. Gates's troops, moreover, were heartened by Bennington and were in no mood to give up more ground. At this point, Burgoyne could have decided to withdraw. He might easily have retired back up Lake Champlain to Canada. But to retreat was not in Burgoyne's nature. He determined to press on, and on 13–14 September his army crossed to the west bank of the Hudson near Saratoga. Burgoyne's optimism was infectious. His officers were still confident. His aide de camp, Lieutenant Sir Francis Carr Clerke, wrote that although the Americans 'pretend to be in spirits and threaten us with a drubbing, ... on the approach of the red Coats I rather believe it will be as usual, they will find out that they can take up better ground in their rear'.[67]

Gates was waiting for the British in a series of defensive works on Bemis Heights which commanded the Hudson and the road from Saratoga to Albany. Elements of the two armies made contact on 18 September: hungry British troops foraging beyond their camp were attacked by American scouts. The following day, Burgoyne sent three columns forward with the aim of turning the American left. Arnold led a strong body of Americans off Bemis Heights to intercept the British advance, and this force clashed with Burgoyne's centre in a clearing in the woods at Freeman's Farm. A bloody battle ensued, with the British centre suffering heavy casualties. Each side displayed great determination: in the words of Colonel John Glover of Massachusetts, 'both armies seem'd determined to conquer or die'.[68] Arnold was unable to persuade Gates to send reinforcements, and at length the Americans disengaged and left the British in possession of the battlefield. Yet this was a Pyrrhic victory if ever there were one. 'Notwithstanding the glory of the day remains on our side', Ensign Thomas Anbury of the Twenty-fourth wrote in his journal, '... the real advantages ... will rest on the that of the Americans'. The British, in Anbury's judgement, were now too weak to continue the advance.[69]

Burgoyne thought otherwise. He was encouraged by the belief that Sir Henry Clinton, left by Howe with around 6,000 men to cover New York, was advancing northwards to assist. On 21 September, Burgoyne received a letter from Clinton promising a push at the American forts on the Hudson about 40 miles above New York City. On 3 October Clinton proceeded with 3,000 troops up the Hudson. On the 5th he took Verplanck's Point without a struggle, and the next day stormed Forts Montgomery and Clinton north of Peekskill. By 7 October the British had captured the hastily-abandoned Fort Constitution, another five miles up-river. From here Clinton sent on 2,000 troops under Major-General John Vaughan to escourt supplies to near Albany, where Burgoyne was supposed to collect them. But on 15 October Vaughan's river pilots refused to go any farther. Bitterly disappointed, Vaughan vented his spleen on the town of Esopus, which was put to the torch after militiamen had 'fired out of the houses' at the British, and because, as Vaughan put it, Esopus was 'a nursery for almost every Villain in the Country'.[70]

Clinton's diversion had fizzled out; but by this time Burgoyne's fate was sealed. On 5 October a British council of war had advised retreat, only for Burgoyne to decide to attack the American positions. Two days later, a reconnaisance force was sent out to probe the American left. When this force was met by a much larger body under Arnold, it gave way. Arnold's men pursued the British back to their camp, overrunning several defensive works in the process. Eventually, the American advance was halted; but not before the British had experienced heavy casualties. Burgoyne now reluctantly agreed to withdraw. At about 11 o'clock at night on 8 October his battered army, less than 6,000 strong, began to retrace its steps. On 12 October Gates succeeded in cutting off Burgoyne's escape route. Unable to recross the Hudson, he had no choice but to open negotiations. Gates, with nearly twice as many troops as Burgoyne, was in a powerful position. Yet he was nervous about Clinton's progress up the Hudson. This nervousness led him to offer generous conditions to hasten Burgoyne's surrender. According to the convention signed by the two commanders at Saratoga, Burgoyne's army would be allowed to march to Boston and embark for Europe, provided it took no further part in the war. Congress, however, repudiated the convention – no doubt fearing that Burgoyne's troops could be used to release an equivalent number for service in America. His army therefore became prisoners, confined first at Cambridge, Massachusetts, and then in Virginia.

The revolutionaries were of course elated by Burgoyne's capitulation. 'It was a glorious sight', Oliver Boardman of Connecticut wrote on 17 October, 'to see the haughty Brittons march out & Surrender their arms to an Army which but a litttle before, they despis'd & called paltroons'.[71] This excitement was felt far beyond the ranks of Gates's victorious troops. The Americans had knocked a British field army out of the war. Never again would they view their enemy with quite such awe. American loyalists were correspondingly depressed. Many held Howe responsible for Burgoyne's defeat, and lamented the earlier opportunities to crush the rebellion which had been missed. The British themselves were also dispirited. From Phialdelphia, Ambrose Serle, Admiral Howe's secretary, recorded on 9 November: 'every thing wears a melancholy aspect; and a Disposition appears very evidently among some principal officers to throw the Blame on others'.[72] But perhaps the most important result of Saratoga was its effect on France. The French government had already been negotiating with the Americans for some time. The French planned to enter the war sooner or later, and might well have done so even without Saratoga. But the British defeat – and Washington's creditable performance at Germantown – certainly encouraged them. When, in December 1777, news of Burgoyne's surrender reached Paris, the French government made a commitment to recognize the United States and to enter into an alliance. The treaties were finally signed in February 1778. From that point, Britain and France, although not technically at war until the following July, were in a state of belligerency.

Chapter 4 Notes

1. Journal of Abner Sanger, 1774–82, Library of Congress.
2. Lt.-Col. James Abercrombie to Lord Amherst, 20 June 1775, Amherst Papers, U 1350 080/13, Kent Archives Office.
3. *Greene Papers*, i. 89–90, 92.
4. Lt. Francis Laye to his father, 12 Oct. 1775, Laye Letters, National Army Museum.
5. Sill to John Spencer, 6 July 1775, Spencer Stanhope Muniments, 60542/8, Sheffield Central Library.
6. Leith Hall Muniments, GD 225/1044, bundle 14, SRO.
7. William Carter, *A Genuine Detail of the Several Engagements ... of the Royal and American Armies ... In a Series of Letters to a Friend* (London, 1784), p.7.
8. Percy to Harvey, 28 July 1775, Percy Papers, Vol. L, Pt. A, fo. 90, Alnwick Castle.
9. Reeve to Sir George Howard, 15 Nov. 1775, Howard-Vyse Deposit, D/HV. B.10/13, Buckinghamshire RO.
10. Lt. Willam Duff to William Rose, 13 June 1775, Duff of Braco Muniments, MS 2226/96/9, Aberdeen University Library.
11. *The Narrative of Colonel Ethan Allen* (New York, 1961 edn.), p.8.
12. Justin H. Smith, *Arnold's March from Cambridge to Quebec* (New York, 1903), Appendix, p.472.

13. *Revolutionary War Journals of Henry Dearbon*, ed. Lloyd A. Brown and Howard H. Peckham (Freeport, N.Y., 1969), pp.49, 56.
14. Harvey to Lt.-Gen. John Irwin, 30 June 1775, WO 3/5, p. 37; Harris to Mrs Dyer, 24 July 1775, Stephen Rumbold Lushington, *The Life and Services of General Lord Harris* (London, 1840), p.58.
15. Timothy Newell, 'A Journal Kept During the Time Y^t Boston was Shut up', Massachusetts Historical Society, *Collections*, 4th series, i (1852), 275.
16. 'The Diary of John Rowe', Massachusetts Historical Society, *Proceedings*, 2nd series, x (1895–6), 97.
17. Gardner to John Radcliffe, 22 June 1776, Radcliffe MSS, D/ER C. 347/3, Hertfordshire RO.
18. Diary of Lt. William Digby, Add. MS 32,413, fos. 18–19.
19. 'Journal of Lieut. Rufus Wheeler', Essex Institute, *Historical Collections*, lxviii (1932), 374–5.
20. 'My Plan for fullfilling the object of the expedition', 14 March 1776, Clinton Papers, William L. Clements Library.
21. 6 July 1776, Dansey Letters, Historical Society of Delaware.
22. Murray to Mrs Smyth, 7 July 1776, *Letters from America*, ed. Eric Robson (Manchester, 1951), p.25.
23. Hart to his father, 7 July 1776, Hart Papers, D 3077/B/1/2, Public Record Office of Northern Ireland.
24. 'Lee Papers', New-York Historical Society, *Collections*, v (1872), 93.
25. Howe to Percy, 9 June 1776, Percy Papers, Vol. L, Pt. B, fo. 13.
26. Capt. Francis Hutcheson to Maj.-Gen. Frederick Haldimand, 10 July 1776, Haldimand Papers, Add. MS 21,680, fo. 122.
27. Isaac Foster Jnr. to his father, 7 July 1776, Isaac Foster Family Papers, Library of Congress.
28. *The New-York Diary of Lieutenant Jabez Fitch*, ed. W.H.W. Sabine (New York, 1954), p.30.
29. Clinton to Charles Mellish, 25 Sept. 1776, Mellish of Hodsock MSS, 171–110/4, Nottingham University Library.
30. Percy's Diary, 1776, Percy Papers, MS 151.
31. Campbell to William Congreve, 9 Nov. 1776, Congreve Papers, S. MS 47/2, William Salt Library.
32. 'Bamford's Diary', *Maryland Historical Magazine*, xxvii (1932), 312.
33. Benjamin Trumbull, 'Journal of the Campaign at New York, 1776–7', Connecticut Historical Society, *Collections*, vii (1899), 196.
34. 'Sergeant John Smith's Diary of 1776', ed. Louise Rau, *Mississippi Valley Historical Review*, xx (1933–4), 259.
35. Capt. John McPherson to William McPherson, 20 Nov. 1776, Macpherson Grant Papers, bundle 707, Ballindalloch Castle.
36. 'Letters of Ebenezer Huntington, 1774–1781', *American Historical Review*, v (1899–1900), 715.
37. *Writings of Washington*, vi. 320.
38. Stephen Collins to Jonathan Amory, 6 Dec. 1776, Stephen Collins Letter-book, 1776–7, Library of Congress.
39. Capt. Thomas Stanley to ..., 23 Dec. 1776, Stanley Letters, MS Am. 228.4, Boston Public Library.

40. Harcourt to Viscount Nuneham, 19 Dec. 1776, *The Harcourt Papers*, ed. W.E. Harcourt (14 vols., Oxford, 1880–1905), xi. 180.
41. *The Pennsylvania Evening Post*, 7 Dec. 1776.
42. *Jefferson Papers*, i. 659.
43. Orderly-book of the Seventeenth Foot, New-York Historical Society.
44. Maitland to his father, 21 Dec. 1776, Maitland Papers, 7902–12–47, National Army Museum.
45. 'A Connecticut Soldier under Washington', ed. William S. Powell, *William & Mary Quarterly*, 3rd series, vi (1949), 102.
46. 'A British Officer's Revolutionary War Journal', ed. S. Sydney Bradford, *Maryland Historical Magazine*, lvi (1961), 167; Sutherland to Dugald Gilchrist, 30 May 1777, Gilchrist of Ospisdale Muniments, GD153, box 1, bundle 4, SRO.
47. Grant to John Grant, 5 June 1777, Seafield Muniments, GD 248, box 509, bundle 3, SRO.
48. *Writings of Washington*, vi. 398.
49. HMC, *Report on the Manuscripts of Mrs. Stopford- Sackville* (2 vols., London, 1904–10), ii. 52–3.
50. 'Letters from a Hessian Mercenary', trans. C.V. Easum and ed. Hans Huth, *Pennsylvania Magazine of History and Biography*, lxii (1938), 498.
51. *Writings of Washington*, viii. 270.
52. Lt-Col. Thomas Sterling to Sir William Sterling, 3 July 1777, Abercairny Muniments, GD 24/1/458, SRO.
53. Stuart to the Earl of Bute, 10 July 1777, in E. Stuart-Wortley, *A Prime Minister and His Son* (London, 1925), p.113.
54. *The Services of Lieut.-Colonel Francis Downman*, ed. F.A. Whinyates (Woolwich, 1898), p.30.
55. Grant to Richard Rigby, 31 Aug. 1777, Macpherson Grant Papers, bundle 772.
56. *Revolution in America: Confidential Letters and Journals ... of Adjutant-General Major Baurmeister*, trans. and ed. Bernhard A. Uhlendorf (New Brunswick, N.J., 1957), p.109.
57. *Citizen Soldier: The Revolutionary War Journal of Joseph Bloomfield*, ed. Mark E. Lender and James Kirby Martin (Newark, N.J., 1982), p.128.
58. Johann Ewald, *Diary of the American War*, ed. and trans. J.P. Tustin (New Haven, Conn., 1979), pp.87, 89.
59. 'Journal of Capt. William Beatty, 1776–1781', *Maryland Historical Magazine*, iii (1908), 110–11.
60. 'Diary of Lieut. James McMichael', *Pennsylvania Archives*, 2nd series, xv (1890), 213.
61. *A Rhode Island Chaplain in the Revolution*, ed. Jeanette D. Black and William Greene Roelker (Port Washington, N.Y., 1972), p.53.
62. *An Eyewitness Account of the American Revolution ... The Journal of J.F. Wasmus*, trans. Helga Doblin, ed. Mary C. Lynn (New York, 1990), p.59.
63. 'Lord Francis Napier's Journal', ed. S. Sydney Bradford, *Maryland Historical Magazine*, lvii (1962), 324–5.
64. *Hadden's Journal and Orderly Books* (Albany, N.Y., 1884), p.95.

65. Schuyler to William Heath, 28 July 1777, 'The Heath Papers', Massachusetts Historical Society, *Collections*, 7th series, iv (1904), 135.
66. Add. MS 32,413, fos. 65, 68.
67. Clerke to Lord Polwarth, 10 Sept. 1777, Polwarth Papers, Lucas Collection, L 30/12/17/14, Bedfordshire RO.
68. Glover to Heath, 21 Sept. 1777, 'The Heath Papers', 156.
69. *With Burgoyne from Quebec*, ed. Sydney Jackman (Toronto, 1963), p.175.
70. Vaughan to [Clinton], 17, 26 Oct. 1777, Clinton Papers.
71. 'Journal of Oliver Boardman', Connecticut Historical Society, *Collections*, vii (1899), 235.
72. *The American Journal of Ambrose Serle*, ed. Edward H. Tatum Jr. (San Marino, Ca., 1940), p.263.

5

Military Operations in America, 1778–83

Saratoga and the Franco-American alliance profoundly changed the course of the war. British officers in America might still breathe defiance – '[we] want you only to give us a few more Men to finish the matter ... a powerful reinforcement, and you have the Country'[1] – but George III and his ministers knew that a different approach was now necessary. Even before the French and Americans signed their treaties, conflict with France – and probably with her ally Spain – was seen as inevitable. A struggle to put down colonial rebellion was about to become a much wider war against the traditional Bourbon foes. Consideration was even given to complete withdrawal from the 13 colonies and the transfer of all available military and naval resources to the impending clash with France and Spain. This drastic option was rejected, but on 8 March 1778 Germain instructed Howe's successor Clinton to prepare for a predominantly naval war to be conducted from bases at New York and Nova Scotia. The abandonment of Philadelphia was recommended, together with a southern expedition to tap reported loyalist strength in Georgia and the Carolinas. The announcement a few days later of the signing of the Franco-American treaties brought forth a further British reappraisal. On 21 March Germain told Clinton that the main effort was now to be directed against France, by means of 'an immediate attack upon the French possessions in the West Indies'.[2] Clinton was ordered to send 5,000 troops against St. Lucia, and another 3,000 to reinforce the Floridas. He was to withdraw the remainder of his army to New York. Peace commissioners, led by the Earl of Carlisle, were to be sent out to open negotiations with the Americans.[3]

*

So Philadelphia was to be given up. In reality, its possession had brought the British little advantage. Excursions from the city which were designed to bring Washington to battle, or to collect much-needed supplies, succeeded only in reducing still further the already-diminished reputation of the British army. 'They did great Damage in their March', recorded John Miller of Germantown after one such excursion, 'Sparing neither Friend nor Foe, burning robing Stealing all the Way they went'.[4] In Philadelphia itself, the British certainly enjoyed better conditions than did Washington's troops at Valley Forge. Even so, not many of the British occupiers would have agreed with Captain John Macleod of the Royal Artillery that 'our Winter Quarters have been exceedingly good'.[5] Although provisions had become more plentiful since the opening of the Delaware, the reduced flow of supplies from the surrounding countryside ensured that prices remained high. For impecunious junior officers like Macleod's regimental colleague First-Lieutenant Francis Laye, the prospect of departure from the city was positively to be welcomed: 'we are all in hopes of Getting away from this Place Soon', he told his father on 12 May, 'for every Necessary of Life is so Extreem^y dear there is no supporting it'.[6]

But for those officers who reflected upon political matters, the evacuation was a bitter blow. Howe's resignation had already dispirited the army – 'His loss is ... universally regretted'[7] - and the idea of abandoning Philadelphia deepened the depression. Sir William Erskine, the quartermaster-general, felt 'miserable in himself & ashamed of the name of a Briton'.[8] Clinton must have been equally dismayed. To be ordered to leave Philadelphia was humiliating enough; to be required to give up a large portion of the army he had just inherited was even worse. With a heavy heart Clinton began preparations for the evacuation, but he decided to delay detachment of the 8,000 troops to the West Indies and the Floridas. He would depart 'with the whole army', he informed Germain on 23 May, and split his forces once he had reached the safety of New York.[9]

Nearly 3,000 loyalists, dreading the return of the revolutionaries, elected to depart with the British. Clinton sent these unfortunates on board ship, together with some of his stores and those troops who were unfit for a gruelling march, and set off with the army proper on the morning of 18 June. Progress was inevitably slow. The British supply waggons alone occupied some 12 miles of road. It took six days to reach Allentown, about 35 miles from Philadelphia. At Allentown Clinton decided to move north-east, towards Monmouth Court House and on to Sandy Hook, in order to avoid a difficult crossing of the Raritan River at New Brunswick.

By this stage, the American army had left Valley Forge and was shadowing Clinton's slow-moving column. Most of Washington's colleagues were so relieved to see the British leave Philadelphia that they were disinclined to pursue them vigorously. The recently-exchanged Charles Lee appears to have been reluctant even to harass the retreating enemy. But Washington was anxious to seize the opportunity to inflict a serious defeat on the vulnerable Clinton. On 27 June, while both armies were trying to recover from the searing heat, Washington ordered the van of his forces under Lee to attack the British once they resumed their march. The next morning, Lee's troops accordingly engaged the British rearguard commanded by Cornwallis. A confused action developed in temperatures near 100° F: 'a Number of our men died with heat' a Rhode Island private noted laconically in his journal.[10] Lee, apparently concerned that Clinton with the main British force was on his way to rescue Cornwallis, decided to pull his corps back. As he retreated, Lee came into contact with Washington and his staff advancing ahead of the rest of the American army. After an ill-tempered interview with Lee, Washington restored order. 'By his own good sense and fortitude', one of Washington's aide-de-camps wrote after the battle, 'he turned the fate of the day'.[11] Forming his troops on high ground, Washington prepared for the British attack. Clinton and Cornwallis, heartened by Lee's retreat, underestimated the strength of the American position. The British launched a series of unco-ordinated and small-scale assaults, all of which were repulsed. At length, they disengaged. Captain Joseph Bloomfield of the Third New Jersey Regiment could hardly contain himself. The Americans, he wrote, had fought 'the flower of the British army' and defeated 'the proud King's-Guards & haughty British-Grenadiers, & gained Immortal-honor'.[12]

In truth, the battle of Monmouth Court House was inconclusive. Clinton had failed to capitalize on the confusion caused by Lee's withdrawal, and Washington had been unable to deliver a crushing blow to the encumbered and exposed British column. On 29 June Clinton's forces continued on their way, without American opposition. On 1 July they reached Sandy Hook, and by the 6th the British were back at New York City.

*

The British fleet moved Clinton's army to New York not a moment too soon. On 11 July the comte d'Estaing arrived off Sandy Hook with 16 French ships of the line with the aim of delivering a knock-out blow to the British forces in North America and bringing the war to a rapid conclusion.[13] This turn of events naturally caused some alarm in the British camp: Admiral Howe's ships were undermanned, and many of the crews of the merchant vessels in port were temporarily pressed to make up the numbers. But in reality the immediate danger to New York was slight: Admiral Howe's numerically weaker force was safe in the harbour and by rapidly occupying Sandy Hook Clinton denied d'Estaing the opportunity to use it as a gun platform from which to shell the British ships. The most obvious impact was on prices in the city. 'The French Fellows have stop^{t} the Fishing Boats', a young officer complained, with the consequence that the cost of fish had soared.[14] Yet, although d'Estaing failed to secure the decisive victory for which the French government had planned, the appearance of the French ships emphasized how the nature of the war had changed. General Howe had been able to conduct land operations confident in the knowledge that his brother's fleet enjoyed naval superiority. Clinton, by contrast, was to be dogged with the worry that the French navy might at any time put his army at risk.

The first real indication of the potential – and limitations – of the Franco-American alliance came not at New York but at Newport, Rhode Island. Its exposed garrison of around 3,000 British, German and provincial troops commanded by General Sir Robert Pigot must have seemed an easy target to Washington and d'Estaing. On 30 July d'Estaing's fleet entered Narragansett Bay. Two days later, General John Sullivan, commanding the American land forces near Newport, declared that 'the enemy are compleatly block'd up'.[15] Within a short time Sullivan had some 10,000 Continentals and militia at his disposal. A co-ordinated Franco-American attack was arranged, but on 9 August Admiral Howe, reinforced by ships from Britain, unexpectedly arrived off Newport. D'Estaing sailed out to do battle, only for a severe gale to prevent the two fleets from effectively engaging. Undeterred by what he hoped was the temporary departure of the French, Sullivan pressed on. On 12 August John Palsgrave Wyllys, a major in the Connecticut Continentals, excitedly told his father that 'Our advance Piquettes are within a mile & an half of their works ... We all have framed the most sanguine expectations of being soon in possession of Newport'.[16] Two days later, an American attack was unsuccessful. Thereafter, Sullivan's militia started to drift home. Damaged French ships began to reappear, but d'Estaing was no longer keen to stay at Rhode Island. Sullivan now openly criticized his allies. D'Estaing was unmoved by American appeals and, fearing that Howe was about to be further reinforced, he set out with his fleet for Boston and its repair yards. As the Americans lifted the siege, complaining about the way in which 'Mr Frenchman Left us ... in the Lurch',[17] the British garrison sallied forth on 29 August. High stone walls separating the fields slowed down the British advance and provided cover for the

Americans. Faced with increasingly stiff resistance, the British halted their attack. The next day Sullivan's force withdrew. After spending so long convinced that they could trap Pigot and his men, the Americans only just escaped themselves. On 1 September British ships brought Clinton and 4,000 men from New York to rescue the Newport garrison.

*

The relief force that arrived at Rhode Island proceeded, under the command of Major-General Charles Grey, to raid the New England coast. New Bedford was burned and large numbers of livestock were seized from Martha's Vineyard. Clinton, belying his reputation for caution, was anxious to do more. While he still commanded the troops intended for the Caribbean and the Floridas, and while the British enjoyed temporary naval superiority, he wanted to destroy the French fleet in harbour at Boston. Admiral Howe refused, on the grounds that the town was too strongly defended. Frustrated in this direction, Clinton turned his attention to Washington's army encamped north of New York City. A British force marched into New Jersey in the hope of bringing Washington to battle. But Washington would not be drawn. The large quantity of forage collected, though much-needed, was scant compensation. October saw another attack on New Jersey; this time further south at Egg Harbor, 'a seminary for little rascally privateers' in the words of Admiral Howe's temporary successor, James Gambier.[18] The local salt works were destroyed, together with shipping. Still more destructive were the raids launched on the north-west frontier of the rebel colonies. In July 1778 a mixed force of loyalists and Indians had devastated Wyoming Valley, Pennsylvania. In November it was the turn of Cherry Valley, New York.

Some British officers doubted the usefulness of the raids – 'these sort of things will never put an end to the war'[19] – but for others who were irritated by their inability to defeat the Americans, destruction, or 'fire and sword' as they usually put it, was the obvious way forward. The raids were seen as presaging a full-scale war of devastation: 'we are in hopes that the fashion may take root', Major Sir James Murray exclaimed after the attack on New Bedford.[20] Widespread destruction, in the view of many officers, would terrorize the rebels into submission. 'Temporising – moderation & delay, have governed our Councils', one complained;[21] if 'the Torch [is] allow'd', observed another, 'I say we'l reduce them still'.[22] Clinton, however, was doubtful – or at any rate was reluctant to engage in such a war. His unwillingness may have been inadvertant wisdom. Those raids that were carried out seem to have done little to further the British cause. They made Major Ebenezer Huntington of Connecticut angry and vengeful rather than submissive. The burning of New Bedford, he wrote, was 'one of the most Wanton Acts of Cruelty they have been Guilty of'. 'It is my wish,' he continued, that, should Grey fall into American hands, 'he should be burnt alive, in a Manner agreeable to the Indian Custom.'[23]

*

In early November 1778 Clinton finally sent his detachments to the West Indies and the Floridas. The loss from his command of 5,000 British regulars he regarded as a particular blow – 'a dismemberment which is severe indeed' he told Germain. But even though he claimed that the departure of so many troops was 'fatal to the hopes of any future vigour in this army',[24] he now chose to act on Germain's earlier suggestion of a southern expedition. Lieutenant-Colonel Archibald Campbell of the Seventy-first Highlanders was sent with about 3,000 men to take Georgia.

Campbell arrived off Tybee Island, 15 miles from Savannah, just before Christmas. General Robert Howe, the American commander in the area, offered some resistance to the British invaders on 29 December, but his small force of Continentals and militia was no match for Campbell's troops. Howe's men surrendered or fled – not just back to Savannah but through the town and beyond. Savannah was left undefended, and the British occupied it the same day. An acute officer, who recognized the importance of winning over the people, Campbell had given strict instructions against plundering. His orders seem to have been obeyed to the letter: all accounts agree that the town and inhabitants were treated with remarkable leniency.

From Savannah, Campbell's progress was somewhat retarded by the lack of accurate maps; but by 3 January 1779 he had reached Ebenezer, where he encouraged the Lutheran parson 'to preach to the Inhabitants against Rebellion and Licentiousness'.[25] Later the same month, Augusta was taken. Reinforced by troops from East Florida under Brigadier-General Augustine Prevost, Campbell was in a strong position. Georgians were flocking in to offer submission and join a new loyal militia. With understandable pride Campbell announced: 'I have got the Country in Arms against the Congress ... I have taken a Stripe and Star from the Rebel flag of America'.[26]

*

Campbell's rejoicing was premature. Although civil government was restored to Georgia in March, the British had not pacified the province. The approach of a reportedly strong body of North Carolina militia led to the abandonment of Augusta. This had the unfortunate effect of leaving the local loyalists exposed to reprisals from the revolutionaries, and so persuaded other loyally-inclined Georgians, as yet undeclared, that caution was still required. Such prudential feelings can hardly have been lessened when 800 Carolina loyalists under James Boyd, who had been marching to join Campbell at Augusta, were surprised and crushingly defeated on 14 February at Kettle Creek. By this stage, Prevost, who was soon to replace Campbell, had already experienced a reverse of his own. Prevost was convinced that the key to the permanent restoration of British control in Georgia was an advance into South Carolina, which was providing the moral and physical support needed to keep the rebellion alive south of the Savannah River. He therefore sent a small detachment to Beaufort, midway between Savannah and Charleston. On 3 February 1779 this detachment was checked by a much larger force of South Carolina militia and obliged to withdraw by sea. A month later Prevost was more successful in an engagement with North Carolina troops at Briar Creek. The defeat of the Americans caused many revolutionary militiamen to return home, but the British position remained precarious. The forces under Benjamin Lincoln, the new American commander in the south, still outnumbered those of Prevost, and in late April Lincoln decided to advance into Georgia.

At this point the British might reasonably have been expected to pull back to protect Savannah. Instead Prevost boldly – Clinton would later say recklessly – launched a counter stroke at South Carolina. His troops met very little resistance as they marched towards Charleston. The advanced guard scattered a small American force about a mile from the town, but the British were themselves too few in number immediately to attack the unprepared defenders: 'Being unsupported & our Numbers not above 500 men: what could we do [?]', a loyalist officer asked plaintively.[27] As soon as Prevost and the main body of British troops arrived, the town was summoned. John Rutledge, the revolutionary governor of South Carolina, was willing to surrender in return for a British acceptance of the neutrality of the state for the rest of the war. Prevost, however, would settle for nothing short of unconditional surrender. News that Lincoln had turned back to relieve Charleston obliged Prevost to lift his siege on 12 May and retire to John's Island. By demanding too much Prevost had ended up empty-handed. But he had at least induced Rutledge to offer terms which exposed the shallow roots of the Revolution in South Carolina. As Congressman Christopher Gadsden remarked: 'we have had a narrow, very narrow, escape indeed, [though] more from the trecherous Whispers and Insinuations of *internal* Enemies than from what our external and open Foes were able to do against us'.[28]

The southern campaigns, 1775–81

Virginia now briefly became the major scene of operations. In April Clinton decided that as the troops opposing Prevost were being reinforced from Virginia an expedition to the Chesapeake would divert American resources from the south. A force of 1,800 men led by Major-General Edward Mathew landed near Portsmouth in May. They proceeded to destroy shipping, naval stores and any materials that might have assisted the American southern army. Their attack on Suffolk was described by John Lawrence, a local merchant, as 'a most Wanton & Villainous affair'.[29] Yet, far from meeting bitter resistance, Mathew's men were hardly obstructed. Indeed, Commodore Sir George Collier, the naval commander of the expedition, reported that 'the most flattering hopes of a return to obedience to their Sovereign may be expected from most of this Province'.[30] Encouraged by this response, Collier wanted to stay and establish a naval base at Portsmouth. But Mathew was unwilling to tarry, and, in accordance with Clinton's instructions, the British sailed away on 24 May.

Back in South Carolina, Prevost continued to defy Lincoln. On 20 June, as the main British force embarked from John's Island, the Americans attacked a covering party of the Seventy-first Foot at Stono Ferry. The Highlanders were heavily outnumbered but well entrenched. They were able to hold back Lincoln's troops and then join their comrades at Beaufort. There the British remained, the intense summer heat sapping the energies of both sides and temporarily bringing an end to the contest. The Americans, however, were determined to extinguish the threat to South Carolina. Apparently unable to accomplish this alone, they turned to the French. In response to a request for help, d'Estaing returned from the West Indies and on 1 September anchored off the coast of Georgia. The admiral had no intention of staying long, but while the hurricane season lasted he was prepared to co-operate with the Americans against Savannah. He landed 5,000 troops on 12 September, and Lincoln's army joined them a few days later. Prevost asked for a day to consider a summons to surrender, allowing time for Lieutenant-Colonel John Maitland to reinforce the garrison with the troops left at Beaufort when Prevost's main force returned from South Carolina. Strengthened by Maitland's men, Prevost now rejected the summons, and formal siege operations commenced. Bad weather caused problems for the investing army: on 22 September Joseph Clay, a Georgia congressman accompanying the American staff, wrote that heavy rain 'has delayed us very much, they [the French] only begun this Day to land the Cannon & Mortars'.[31] The time lost enabled the British to improve their defences. On 5 October the besiegers started to shell the town. Even after three days' bombardment Prevost refused to yield. D'Estaing's engineers told him that the siege could be a protracted one. He was unwilling to wait. On board his ships, scurvy and dysentery were making deadly inroads into the crews; and, according to a French officer, 'All the vessels were damaged. Most of them had their rudders in a piteous condition, having suffered much from gusts of wind and strong sea.... The siege lasting for some time yet, it was feared that many ships would be put out of service'.[32] If Savannah could not rapidly be pounded into submission, then it would have to be stormed. Almost all of his colleagues opposed this risky course, but d'Estaing was adamant. The next day the allied army attacked. The British, warned of the impending assault by a deserter, were ready and waiting. The attackers were exposed to a withering fire and were repulsed with dreadful casualties: 900 French and Americans were killed or injured, compared with only 54 members of Prevost's garrison. Lincoln wanted to resume formal siege operations, but d'Estaing was unwilling to remain in Georgia. The siege was abandoned on 18 October. The promise of successful Franco-American co-operation, disappointed at Newport the year before, remained unfulfilled.

For the time being, at least, it was the Spanish who were more effectively to bring their power to bear in North America. Spain entered the war in June 1779, though as an ally of France rather than the United States. The loss of Florida to Britain at the end of the Seven Years' War still rankled at Madrid; its recovery was accordingly a high priority. Bernardo de Gálvez, the governor of Louisiana, acted quickly and found the British in West Florida unprepared. In September his forces took the posts at Manchac, Baton Rouge and Natchez. As one of Gálvez's elated subordinates put it: 'Your Lordship with the loss of only one man has obtained the surrender of 400'.[33] The embarrassed British commander was obliged to confess 'the conquest of the western part of this province by the arms of Spain in consequence of their early intelligence of the commencement of hostilities'.[34]

*

In 1779 Britain's enemies were successful in another area remote from the main theatre of operations. The highly-destructive Indian raids of 1778 persuaded Washington that something had to be done to combat this menace to the back-settlements of New York and Pennsylvania. Sullivan was sent with 2,500 Continentals to vanquish the Iroquois, and was joined by further troops from New York. On 29 August Sullivan's army routed a mixed force of Indians and loyalist rangers at the village of Newtown. Unaccustomed to an artillery attack, the Indians fled: they were, as Sergeant Moses Fellows of the Third New Hampshire Regiment recalled, 'most Horridly Pannic Struck'.[35] After further skirmishing, the Indians retreated westwards towards the British post at Niagara, leaving the Americans free to wreak terrible damage on their lands. Sullivan had been told by Washington to destroy the Indian settlements and their means of subsistence, and he proceeded to lay waste the Mohawk Valley and the Genesee Country. This systematic devastation had the effect of compelling the Iroquois to become a burden on the British; but Washington's most ambitious expectations were not met. Sullivan was unable to push on and take Niagara. Nor were the Indians knocked out of the war. In 1780 and 1781 they returned to terrorize the New York frontier.

At New York City, meanwhile, Clinton still hoped to bring the elusive Washington to a decisive action. The British commander reckoned that an advance on the Highlands astride the Hudson to the north of the British lines might draw the main American army from its quarters in New Jersey. Stony Point and Verplanck's Point were seized with little difficulty on 1 June, threatening Washington's communications with New England. Washington moved to cover the new defences at West Point, further up the Hudson, but he refused to come out and fight. The game of cat and mouse continued when Clinton sent Major-General Tryon to raid the Connecticut coast. Entrusting the task to Tryon was perhaps unfortunate. An enthusiastic advocate of 'a Depredatory War',[36] he proceeded to exceed the letter and spirit of his orders, much to Clinton's discomfort. New Haven, Fairfield and Norwalk were all burned between 5 and 11 July, and at the same time Tryon's troops added to the havoc by seizing moveable property and attacking civilians. The inevitable protests followed about the British 'wageing War against innocent Women & Children'.[37] Ezra Stiles, president of Yale, was just one of many outraged Americans: 'Their Barbarity was savage & cruel', he wrote in his diary of the British raiders, '... to the degree of the highest & most brutal Rigors of War'.[38] Washington's response was more measured. He sent troops to Connecticut, but rather than move his main force into New England and expose himself to attack by Clinton's army – as Clinton fervently hoped that he would do – Washington on 15 July launched a surprise assault on the posts in the Highlands recently taken by Clinton. While the garrison at Verplanck's held out, Anthony Wayne's light infantry took Stony Point by storm. Tryon's troops were immediately recalled by Clinton; but Washington, having achieved his objective, destroyed the works at Stony Point and withdrew.

Further north, the British were more successful. Earlier in the year Clinton had been instructed to establish a settlement for loyalist refugees at Penobscot Bay, then in Massachusetts, now in Maine. Some 640 troops from Halifax disembarked on 16 June. Their commander was Brigadier-

General Francis McLean, who was described by a perceptive Hessian commentator as 'probably the first Briton who understands the art of winning the confidence of the inhabitants'.[39] As in Georgia, so at Penobscot: local people flocked in – 'about five hundred have ... taken the Oath of allegiance & fidelity', McLean proudly told Clinton ten days after the British had landed.[40] The revoltionary authorities of Massachusetts, however, were not inclined to leave McLean undisturbed. A force of Massachusetts militia and state and Continental warships sent to dislodge the British reached Penobscot on 24 July. The Americans began siege operations and the local people, as McLean perhaps over-generously put it, 'were compelled to join the Enemy'.[41] The British post, though by no means strong, held out long enough for a naval squadron under Collier to come to the rescue. On Collier's arrival, the besiegers panicked. Rather than fight, all but two of Commodore Dudley Saltonstall's ships tried to seek shelter up the Penobscot River. There they ran aground and had to be set alight by their crews. The sailors and militia fled into the woods. The ever-patient McLean then began afresh the task of winning over the inhabitants.

All this was of little consequence to poor Clinton. For him, Penobscot constituted no more than a side-show – a distraction from the war proper. He still resented the loss of 8,000 men to the West Indies and the Floridas, and another 2,000 to reinforce Canada (thought to be a likely French target). Worse still, he was frustrated by his inability to persuade Washington to fight the battle that the British so desperately wanted. Clinton had already offered his resignation in 1778, just before he detached a sizeable chunk of his army to other theatres, and when on 19 August 1779 the Americans daringly seized Paulus Hook – just across the Hudson from New York City – he asked again to be replaced. His request, as in 1778, was turned down, and Clinton remained a disgruntled and frequently dejected British commander-in-chief.

His subordinates seem to have been similarly downcast. But some of them, rather than blame shortage of troops and neglect by the government – Clinton's almost constant refrain – pointed the finger at Clinton himself. John Mervin Nooth, an army surgeon, was one of the malcontents. 'Nothing surely can be more shameful than our perfect Inactivity', he complained, 'it is unfortunately our fate to be commanded by a Person that has no Abilities to plan nor Firmness to execute the most trivial military Operations'.[42] Clinton's decision to evacuate the Highlands and the exposed garrison at Newport and concentrate his forces at New York – a response to the appearance of the French navy off Savannah – no doubt contributed to Nooth's rage. But retrograde as these steps appeared, they had the virtue of giving Clinton sufficient troops to spare for a significant field army. Despairing of victory in the north, Clinton was now prepared to commit significant forces to Germain's southern strategy: if South Carolina could be pacified, the British might then advance steadily northwards, reclaiming the insurgent colonies one by one.

*

On 26 December 1779 Clinton set sail with 7,600 troops to take Charleston. Severe storms scattered the British ships and made the voyage wretched for soldiers and sailors alike. Not until the end of January 1780 did most of the transports reach Tybee Island, near Savannah. They were, of course, well south of their original objective; but after such a battering repairs needed to be carried out and the troops welcomed the chance to recuperate. On 10 February the fleet sailed again and reached North Edisto Inlet, about 30 miles from Charleston, the next day. Lincoln made no attempt to block the advance of the disembarked British troops; nor did he seek to withdraw his own men from Charleston. He just sat and waited, expecting reinforcements from the north to strengthen his hand. Political considerations influenced him, too. To risk his outnumbered army in battle could put the whole of the South in jeopardy, while to abandon Charleston – the largest town in the region – might be equally damaging to American morale.

Moving slowly across the network of islands and waterways along the coast, on 6 March Clinton's troops were able to occupy Fort Johnson on the southern shore of Charleston harbour. Two weeks later Admiral Arbuthnot's frigates negotiated the sand bar and crossed into the lower harbour. The American naval flotilla in the upper harbour was thereupon scuttled to block the entry of Arbuthnot's ships into the Cooper River. But the main British effort came from a different direction. At the end of March detachments of Clinton's army crossed the Ashley River a dozen miles above Charleston. On the night of 1–2 April the first siege trenches were dug on Charleston Neck, to the north of the city between the Ashley and Cooper Rivers. A few days later the arrival of 700 Continental troops cheered the garrison; but this force was insufficient to save the city and in the end merely swelled the number of American prisoners. Clinton's artillery began firing on 13 April, and the next day Lincoln's escape route was finally cut off when Lieutenant-Colonel Banastre Tarleton of the loyalist British Legion seized Monck's Corner, a strategically important post on the Cooper River. On 21 April Lincoln offered to surrender on condition that he and his troops were allowed to march out of the city as free men. Clinton was in no mood to be so generous. The noose tightened. By the first days of May British entrenchments on the Neck were only yards away from the American defences. Clinton was anxious to avoid storming the city: this might be too bloody a prelude to what he hoped would be the reconciliation of the southern colonists. He was therefore growing increasingly exasperated with the stubbornness of the garrison. 'I begin to think these People will be Blockheads enough to wait the Assault – Je m'en lave les Mains', he wrote on 6 May.[43] But Lincoln recognized the futility of continued resistance. He opened negotiations again, only for Clinton to reject his terms. On 9 May the bombardment recommenced. 'The fire now upon both sides was really dreadful', one of the defenders noted in his diary, 'and within our lines did great execution'.[44] The citizens pressed Lincoln to surrender, and on 12 May Charleston capitulated. More than 2,500 Continentals were made prisoners, and about 800 militia were paroled.

Even on the verge of success there were cautionary British voices: 'I am sorry to say when we have got Charles Town we have not got America', wrote Lieutenant-Colonel Thomas Dundas of the Eightieth Foot.[45] But most of the besiegers greeted Lincoln's surrender with great excitement. Clinton himself, flushed with triumph, anticipated the rapid return to British rule of both the Carolinas. Not even news of the fall of Mobile, in West Florida, to Gálvez's all-conquering Spaniards diminished his satisfaction. His reply to the British commander in West Florida wasted little space on commiserations: most of Clinton's letter comprised an account of his own success at Charleston. In the immediate aftermath of the surrender he had every reason to be confident. On 29 May Tarleton caught and defeated a body of Virginia Continentals which had failed to reach Charleston in time and was retreating northwards. This victory at the Waxhaws suggested that the British hold on the tidewater could with comparative ease be extended into the back settlements. On 8 June Clinton sailed to New York with 4,000 men to parry a reported French threat to its denuded garrison. Cornwallis was left in charge of operations in the south. To a general who had long wanted an independent command, the prospects must have seemed bright. From Camden one of his officers wrote home on 10 June: 'This Country is intirely conquered; the People crowd in from all quarters to deliver up their arms'.[46]

*

Once again, the British had been enchanted by a false dawn. Within a few weeks the backcountry had burst into flame. As early as the middle of June, isolated soldiers were being attacked; in July and August a series of clashes took place between revolutionaries and loyalists; throughout the summer British detachments in the back settlements were faced with a rising tide of violence and rebellion. Almost-fulfilled hopes of a return to tranquility in South Carolina were dashed.

Robert Gray, a local loyalist, was later to argue that the core of the problem was 'the disaffection of the Whigs', by which he meant the continued military resistance of small groups of intransigents who refused to accept British victory.[47] By carrying on the struggle, these malcontents provided rallying points for all those driven into opposition to the British. But why did so many inhabitants of the backcountry join in this 'Second Revolution'? The plundering, assaults and rapes committed by members of the army on South Carolinians almost certainly played a part; Tarleton's British Legion gained a special notoriety in this respect. The newly-formed loyalist militia were perhaps even more guilty, using their position of power to settle old scores. Just moving troops into the backcountry might have been a mistake, regardless of the conduct of the soldiers and those who acted under their protection. The area had already been largely stripped of foodstuffs as a result of the demands of the American army at Charleston before the siege. The consumption of the advancing British only exacerbated the shortages and caused friction with local people. The arrival of the troops seems also to have drawn the British into ethnic rivalries that pre-dated the war. The backcountry Scotch Irish were already unfriendly with the local German community. When the Germans started to express their loyalty and ingratiated themselves with the army, the Scotch Irish almost inevitably gravitated towards the revolutionaries. Wild rumours added to British difficulties. A story that the army seized all young men and sent them into Hessian service caused particular alarm: 'it is inconceivable the Damage such Reports has done', Lieutenant-Colonel George Turnbull told Cornwallis.[48] But the reality of British policy was probably still more damaging. On 3 June Clinton had issued a proclamation revoking the paroles given to all those South Carolinians who had submitted. From 20 June they were obliged to choose between actively supporting British measures and being regarded as open enemies. The proclamation, many of Clinton's subordinates complained, made quiet neutrality impossible. Americans who were not well-disposed to the British but had accepted the futility of continued resistance, were forced into armed opposition: 'nine out of ten of them are now embodied on the part of the Rebels', Lord Rawdon wrote from Camden on 7 July.[49]

The decisive ingredient in this dangerous cocktail, however, was the backcountry perception that the Revolution was alive and well in North Carolina. Significantly, in the tidewater area, which also experienced the distresses associated with a British military presence, there was no comparable uprising. The British were too numerous and would-be insurgents were too remote from external support. In the backcountry, by contrast, resistance could more easily be sustained from the neighbouring state. As Turnbull noted on 16 June, the revolutionary militia across the border 'over awes great part of the Country and Keeps the Candle of Rebellion still Burning'.[50] On 20 June a premature rising by North Carolina loyalists was crushed. The British position now appeared still less secure in the South Carolina backcountry, especially as news spread of the advance of a new American army under Horatio Gates, the victor of Saratoga.

On 16 August Cornwallis, although outnumbered, succeeded in removing this immediate threat at the battle of Camden. Gates made the error of entrusting his left wing to militia from North Carolina and Virginia, unsupported by Continental troops. Cornwallis's men broke the militia with ease, while the Continentals on Gates's right held their own and even counter-attacked. For half and hour, in the words of Captain Robert Kirkwood of the Delaware Continentals, the fighting was 'very Desperate'.[51] The Americans crumbled, however, when Tarleton's cavalry outflanked them. Cornwallis's triumph might have been expected to restore tranquility. Indeed, according to Colonel Otho Williams of the Sixth Maryland Regiment, 'the Inhabitants ... were immediately in arms against us' and escaping Continental officers 'were disarm'd by those faithless Villains who had flatter'd us with promises of joining us against the Enemy'.[52] But, in fact, the impact of Camden was distinctly limited. Beyond the immediate environs of the battlefield, there was no significant change in local opinion. Persistent reports of the immmient arrival of further American troops – or even the French – kept the rebellion alive in the backcountry, as did the failure of news of the British victory to penetrate far into those areas in revolt: 'any person daring to speak of it being threatened with instant death', Cornwallis complained.[53]

The only solution was assumed to be an invasion of North Carolina. An advance into this state had always been Cornwallis's intention, but after South Carolina had been pacified, not before. He now believed that until the back settlements of South Carolina were denied physical and moral aid from their northern neighbour, there could be no local perception of British predominance so necessary to produce peace. Besides, the 'friends to government' in North Carolina were thought to be numerous, and they seemed to offer a better prospect than the irresolute South Carolina loyalists. On 8 September Cornwallis started to advance up the Wateree River from Camden. He reached Charlotte in North Carolina on the 26th, where he waited to give his army – a large proportion of which was sick – chance to rest and recuperate. Further west, and also advancing towards Charlotte, was Major Patrick Ferguson and a body of loyalist provincials and militia. Ferguson hoped to recruit more loyalists as he progressed, but all he encountered was fierce opposition from the local people. Harassed by the backwoodsmen, on 6 October he took up a defensive position on King's Mountain. The next day his force was overwhelmed; Ferguson and many of his command were killed, the rest captured. Several of the loyalist militia officers were then executed, 'a sacrifice', wrote Ferguson's adjutant, 'to their infamous mock jury'.[54]

King's Mountain led Clinton to mutter about the stupidity of leaving detachments unsupported, and so caused the relationship between Cornwallis and Clinton to deteriorate (Clinton had already reacted touchily to rumours during the siege of Charleston that some of his officers would prefer to see Cornwallis in charge). Of more immediate importance was the battle's impact on local sentiment. Loyalists were dismayed and frightened, while the spirits of the revolutionaries were raised – increasing still more the threat to the British posts in the backcountry.

King's Mountain also persuaded Cornwallis that he needed reinforcements. He ordered Brigadier-General Alexander Leslie to bring his men from the Chesapeake. Leslie had been sent to Virginia by Clinton in October, partly at Cornwallis's request and partly because Clinton himself saw the area as strategically important. Leslie's instructions were in some respects similar to those Mathew had received in 1779: destroy stores and provide a diversion to help British forces further south. But, unlike Mathew, he was ordered to establish a base and to regard himself as part of the southern army. Leslie's troops, despite his best endeavours, indulged in the same kind of irregularities as had Mathew's. But, as in 1779, British indiscipline failed to rouse the Virginians to resistance. Leslie's force remained in the area around Portsmouth and Suffolk totally unopposed. The apparent acquiesence, even loyalty, of the inhabitants – which seemed to contrast so sharply with their disposition in South Carolina – strengthened the reputation of the area as one suitable for British operations. It also made departure painful for those of Leslie's officers who feared the consequences for the loyalists who had declared themselves and were now to be left unprotected. 'With the heaviest Heart I ever felt I sit down to inform you that we are embarking' was how one such officer described his feelings.[55]

By this time, Cornwallis's army had trudged back to Winnsboro, South Carolina. His retreat was another consequence of King's Mountain. He considered a junction with Leslie in North Carolina, but in the end Leslie went to Charleston and from there marched to join Cornwallis in the new year. At Winnsboro the British spent a miserable winter, short of supplies and depressed by news of setbacks in skirmishes with local partisans and revolutionary militia. The war in the backcountry, always nasty, had degenerated into a particularly savage and bloody affair. Nathanael Greene, the newly-appointed American commander in the south, reported on 29 December: 'The Whigs and Tories pursue One another with the most relentless Fury killing and distroying each Other wherever they meet. Indeed a great Part of this Country is already laid waste & in the utmost Danger of becoming a Desert'.[56] No wonder Cornwallis planned to resume his advance into North Carolina as soon as possible; apart from loyalists in abundance, it offered an escape from the anarchy that was the South Carolina backcountry.

*

In January 1780, while Clinton was sailing south to Charleston, American troops caused a momentary alarm by landing on British-held Staten Island. But besides this bold stroke, and continuing British and Indian attacks on the New York frontier culminating in the capture of Fort George in October, in 1780 the war in the northern colonies was for both sides a weary tale of unfulfilled plans and disappointed expectations.Washington's army had been through a hard winter at Morristown, New Jersey, and his men were undernourished and discontented. Washington himself was under pressure to detach troops to the south. He was therefore forced to remain on the defensive, despite his wish to attack New York. The French expeditionary force which arrived at Rhode Island in July seemed to provide the reinforcements Washington so desperately needed; but the French commander, the comte de Rochambeau, was opposed to close co-operation. When Washington urged the joining together of the armies, Rochambeau pointed out that he was under instructions to remain with the French navy, which had established itself at Newport. Not even the idea of a winter campaign against Canada could entice Rochambeau, whose troops settled into quarters at Rhode Island after a non-too-friendly welcome ('the local people', a French artillery lieutenant wrote, 'would have preferred ..., I think, to see their enemies arrive rather than their allies. We inspired the greatest terror in them').[57]

For the British at New York, 1780 was equally unproductive. In late June an advance into New Jersey was opposed by the local militia and resulted in the burning of Springfield. It was 'not to be wondered at', a subaltern in the Coldstream Guards exclaimed, 'that the soldiers should have wished to fire the houses from which the rebels had fired on them'.[58] This incursion, however – like so many before – failed to produce a battle with Washington's army. Soon after, Clinton, who had returned from South Carolina when he feared that New York might be threatened by the

French, proposed to attack Rochambeau's troops as they found their land-legs at Rhode Island. Admiral Arbuthnot, lethargic and cantankerous as ever, found reasons for delay; and Washington, by advancing on Kingsbridge, on the northern edge of New York's defences, caused Clinton to think again. On 14 September Admiral Sir George Brydges Rodney arrived at New York with 10 ships of the line. He was keen to take on the French in concert with Clinton. This time Clinton put barriers in the way: the French had already fortified Rhode Island to such an extent that an attack was no longer practicable with the small number of troops at his disposal. Clinton's attention, in fact, was focused once more on the strategically important Highland forts above New York City.

Clinton's hopes centred on Benedict Arnold, the real victor of Saratoga and now commandant of West Point. Arnold had become embittered by what he saw as shabby treatment by Congress. He disapproved of the alliance with France, and his marriage to the daughter of a loyally-inclined Philadelphian family probably further sapped his commitment to the revolutionary cause. Large personal debts may also have predisposed Arnold to be receptive to British approaches. In return for £20,000 he agreed to betray West Point to Clinton. The plot was uncovered, however, when John André, Clinton's adjutant-general, was captured in civilian clothes after a meeting with Arnold. Arnold escaped, and André was hanged as a spy.

Arnold's behaviour caused great dismay and uneasiness among the revolutionaries; his name became synonomous with treason of the worst kind. The British mourned the death of André: 'there never was so Melancholy a thing', wrote Major Harry Rooke, 'It is not an Easy matter to describe how much the whole Army is shocked'.[59] They also regretted another lost opportunity. As Rodney remarked of West Point, 'The rebels look upon it as their *dernier ressort*, and would have been undone had our scheme with Arnold succeeded.'[60] Arnold, for his part, became a brigadier in British service, and as such sailed from New York at the end of the year as leader of yet another expedition to Virginia.

*

1781 was to be the decisive year of the war in America, though in the north it started for the revolutionaries in a distinctly unpromising manner. Washington's army was in an even more deplorable state than 12 months earlier. In January lack of pay and shortages of food led regiments from Pennsylvania and New Jersey to mutiny. The next month Massachusetts and New Jersey troops clashed in a serious riot at Princeton. The French, made wary by the apparent collapse of the American army, even considered evacuating their forces from Newport. Washington sent some of his troops to Virginia under the marquis de Lafayette, a French officer in the American service; but until Rochambeau agreed to close co-operation in late May, the main army of Continental troops was in no position to threaten Clinton or to influence the outcome of the war elsewhere.

In the far South, Gálvez continued his triumphant progress through West Florida. The Spanish started to seal off Pensacola in March, their ships entering the harbour in the middle of the month. On the 22nd Gálvez was reinforced, and on 19 April further ships and troops arrived from Havannah. The Franco-Spanish besiegers now amounted to more than 7,000 men. Major-General John Campbell, the British commander, resisted as long as possible; but on the morning of 8 May, in the words of a Spanish account, when the British resumed firing, 'we replied with two howitzers ... with such success that one of our grenades having fired the [British] powder magazine it blew up'.[61] The explosion killed many of the defenders, and in the afternoon Campbell ran up the white flag. In his official report, Campbell justifiably complained of the lack of naval support. But to the British West Florida was always a side-show, a low priority when it came to the allocation of sparce resources.

Even the Carolinas and the Chesapeake seemed to be of secondary importance to Clinton. He had been nervous about the security of New York ever since the French entered the conflict. When he learned that Admiral de Grasse had left Europe with 20 ships of the line, he became still more worried. Rather than sending troops to the South – as Washington had done – Clinton through the spring and summer kept asking for a detachment from Cornwallis's army – first to help defend New York, and then for a projected (but in the end aborted) attack on Philadelphia. But despite Clinton's preoccupation with the middle colonies, the struggle was to be decided further south.

*

In the first week of January 1781 Cornwallis left Winnsboro and headed towards North Carolina. His army was small – only around 1,300 men – but he expected to be joined soon by Leslie with a further 2,000. A detachment was sent from Charleston to Wilmington, on the estuary of the Cape Fear River, to establish a base and gather supplies for Cornwallis's troops. Lord Rawdon was left in South Carolina with nearly 5,000 men to cover Charleston, Savannah and a string of isolated posts inland.

At Charlotte, Greene had made the bold decision to divide his forces.

He sent about 700 Continentals and militia under Daniel Morgan to the west, to collect much-needed provisions and probe the British defences in the South Carolina backcountry. Another body, under Henry Lee, he despatched to co-operate with the partisan leader Francis Marion in attacks on the British positions along the South Carolina coast. The remainder of Greene's army moved to Cheraw on the Pee Dee River.

Cornwallis could not resist the temptation to destroy part of Greene's army. He sent Tarleton with 1,100 troops to deal with Morgan. Tarleton travelled fast – as usual – and met the Americans at Hannah's Cowpens on 17 January. Tarleton's career as a light cavalry officer had been one of almost uninterrupted success, and before leaving Winnsboro he had hoped 'my good Genius still smiles'.[62] But his luck – and judgement – deserted him. A loyalist officer in his force ascribed the British defeat to 'some dreadful bad management';[63] and Tarelton was certainly impetuous. But Morgan deserves credit. By skillful use of the militia, then by pretending to withdraw only to counter-attack, he was able to catch the British off-balance. Tarleton's men were routed; over 800 being killed or captured.

Cowpens was as disastrous for the British as King's Mountain – perhaps more so, for Tarleton's force contained British regulars and loyalist provincial light troops whose loss was a grievous blow. Yet Cornwallis was determined to press on. After failing to catch Morgan, who moved to rejoin Greene, Cornwallis forded the Catawba River on 1 February, brushing aside the North Carolina militia who tried to contest his crossing. Cornwallis now closely pursued Greene all the way to the Dan (or Roanoke) River. The Americans narrowly won this so-called Race to the Dan, crossing into Virginia on 13 February. Cornwallis declined to follow. His men were hungry and tired after a gruelling march through a country desperately short of provisions, and he was mindful of the need to rally the North Carolina loyalists. He therefore pulled back to Hillsborough and called on the people to join him. The response was poor; not least because a body of loyalists under Colonel John Pyle that did attempt to reach Hillsborough was cut to pieces by American cavalry. Brigadier-General Charles O'Hara recalled sardonically that 'the novelty of a Camp in the back Woods of America, more than any other cause, brought several People to stare at us, their curiosity once satisfied, they returned to their Homes ... we never had with us at any One time One Hundred Men [i.e. loyalists] in Arms'.[64]

When Greene recrossed the Dan, Cornwallis left Hillsborough on 25 February. Greene successfully avoided battle until he was reinforced, and at length, on 15 March, the two sides made contact near Guilford Court House. Greene outnumbered Cornwallis's bedraggled army by more than two to one. None the less, the American general resolved on defence, drawing up his men in three lines. The first, composed of militia, fired at the advancing British and then fled. The second, also militia, offered stiffer resistance before falling back; while the third, made up of Continentals, counter-attacked. To halt their advance, Cornwallis took the drastic step of ordering his artillery to fire into the confused mass of struggling British and Americans troops. As the two sides disengaged, and the British reformed, Greene decided to retreat. 'We took the enemy's cannon (four pieces) and consequently a victory, though I must own, without any very brilliant advantages arising from it.'[65] The writer, a subaltern in the guards, understated the hollowness of Cornwallis's triumph. This was a Pyrrhic victory to rank alongside Bunker Hill and Freeman's Farm. Greene lost half as many men as Cornwallis, who was left with an effective force only 1,300 strong. Captain von Closen of the French Deux-Ponts Regiment only slightly exaggerated when he observed: 'The Americans lose 600 men in a day, and 8 days later, 1,200 others rejoin the army; whereas to replace 10 men in the English army is quite an undertaking.'[66]

Cornwallis again appealed for North Carolina's 'friends to government' to come forward; but the loyalists were understandably reluctant to seek the protection of such a palpably weak British force. Disappointed – disillusioned – Cornwallis's 'army' limped to the shelter of British-held Wilmington.

*

While Cornwallis's troops were recuperating, Greene decided to march into South Carolina. He felt confident that he had seized the initiative. 'One of two things must happen', he told Thomas Jefferson on 6 April, 'they must either give up their posts in South Carolina or return to support them; in either case our purpose will be answerd, as it will be undoing all they have done the whole campaign'.[67] Cornwallis left his options open: he wanted to join the British force already operating in Virginia; but he was ready to embark for Charleston if Lord Rawdon found Greene too much to handle.Greene's army advanced on the British base at Camden. Rather than wait to be attacked, on 25 April Rawdon led the Camden garrison against the American camp at Hobkirk's Hill. The struggle was a fierce one. Eventually, however, Rawdon's outnumbered force prevailed. The Americans fell back and were pursued for some miles. Josiah Martin, the last royal governor of North Carolina, described Rawdon's success as 'one of the first achievements of this war'.[68] But in strategic terms the second battle of Camden was no more productive for the British than the first had been. Guilford Dudley, a North Carolina militia officer, recalled that the action 'terminated unfavourably to the American army, though without affording the least advantage to Lord Rawdon'.[69] The British evacuated Camden on 10 May. On the same day Orangeburg fell to revolutionary partisans, then Fort Mott on 11 May, and Augusta on 5 June. 'The Revolt is very general', Rawdon reported bleakly to Clinton.[70] In all of South Carolina and Georgia, only Charleston, Savannah and the remote post at Ninety-Six remained in British hands. Greene was determined to clear the British from the backcountry before turning his attention to Charleston, and Ninety-Six was placed under close siege in June. Again Greene was thwarted; and again the British were unable to benefit from their success. Rawdon, reinforced by three regiments from Ireland, marched to the relief of Ninety-Six, only to abandon the post shortly afterwards.

Rawdon, who fell ill on the march back to the coast, was replaced as commander of the British forces by Lieutenant-Colonel Alexander Stewart. Stewart was content to guard the approaches to Charleston; but Green was sufficiently confident to want to attack. On 8 September the American army moved towards the British camp at Eutaw Springs. About three miles from the British post, Greene's troops came across a foraging party. The firing alerted Stewart, and the British advanced to meet the approaching Americans. Greene's hard-bitten Continentals halted Stewart's men and then pushed them back in disarray. Yet even now victory eluded the luckless Greene. On the British right, Major John Marjoribanks and the grenadiers and light infantry offered stubborn resistance, while other British troops, occupying a house in the camp, poured fire on the jubilant Americans. But the real problem was the indiscipline of Greene's men. On entering the British camp, they broke ranks to plunder the tents and baggage. The momentum of the advance was lost; the British rallied; and Greene was compelled to withdraw.

As at Guilford Court House and Hobkirk's Hill, however, Greene's battlefield disappointment did little to cheer his opponents. Stewart's losses obliged him to pull back to Charleston Neck. The war in South Carolina and Georgia was all but over.

*

Cornwallis set off from Wilmington on 25 April. When shortly afterwards he learned of Rawdon's success at Hobkirk's Hill he felt free to move in earnest towards Virginia. His force – less than 1,500 strong – reached Petersburg on 20 May. By this time, the British had been operating in the Chesapeake for some five months. Arnold had been sent by Clinton to destroy stores, create a diversion in favour of Cornwallis, and, in belated response to Sir George Collier's advice in 1779, to establish a naval base. In a foretaste of difficulties to come, in January 1781 Arnold found his position threatened by three French ships of the line that had slipped past the British fleet watching Rhode Island. In March the whole French naval force at Newport was able to escape Arbuthnot's ships and head for the Chesapeake. An indecisive action took place off the Virginia Capes on 16 March. The French gained the upper hand but declined to press their advantage. Lafayette, who anticipated co-operating with the French fleet against Arnold, was bitterly disappointed. Unaided, his small force could do little to hinder the British.

Later the same month Major-General William Phillips arrived with 2,000 more men and succeeded Arnold as British commander in the area. His army continued its destructive course, using gunboats carrying from 50 to 80 men to reach the shallow parts of the river system. These incursions met with scant resistance from the local militia. Benjamin Gilbert, a New England soldier in Lafayette's force, was disgusted by the supineness of the Virginians: 'it is as hard a matter to find a sincere friend to his Country in this part of the State as to find a Tory in the State of Masachts Bay'.[71] John Bannister, a Virginian officer, was understandably more charitable: 'the People', he explained, 'are tired of the War & come to the Field most reluctantly'.[72]

In late May, then, Cornwallis's junction with the British forces already in Virginia gave him command of an army of more than 7,000 men. He decided to make the destruction of Lafayette's detachment his chief objective. The British forced the marquis to abandon his base at Richmond; but, failing to catch him, they then withdrew to Williamsburg. Lafayette rashly followed Cornwallis. On 6 July the British almost trapped him. Lafayette's advance guard under Wayne ('anxious to perform wonders' as one American officer disparagingly remarked)[73] only narrowly escaped a crushing defeat. Wayne's men, finding themselves facing the whole British army rather than the rearguard they had expected, 'were obliged to retreat with precipitation'. 'Happy for us', Lieutenant William McDowell of the First Pennsylvania Regiment wrote, 'the enemy did not press us at this crucuial moment, or our troops would have inevitably been cut off'.[74]

Cornwallis was now instructed to fortify a base for the use of the navy as a winter anchorage. Portsmouth had already been rejected, and of the other options Old Point Comfort was thought to be too difficult to defend. Yorktown, which could accommodate ships of the line, was Cornwallis's choice; and early in August his troops started to build defensive works. In retrospect, we can see that this was a fatal move. Mobile and on the offensive, the British army constituted a formidable threat. On the defensive in a fixed post, it became a vulnerable target.

The French and Americans were already concerting plans to concentrate their resources against Cornwallis. On 10 June Rochambeau's army had left Rhode Island to meet Washington at Phillipsburg, New York. A planned attack on the British lines came to nothing; but on 19 August the allied army started to move south. Admiral de Barras sailed from Newport a few days later with the siege guns to rendezvous with de Grasse's fleet coming up from the West Indies with 3,000 more French troops. Even at this stage Clinton was uncertain about his enemy's intentions. Washington and Rochambeau did all they could to keep him guessing. For some days they adopted a route that suggested an attack on Staten Island or Sandy Hook might be in the offing. 'General Clinton was completely duped and that was what we wanted', wrote one of Rochambeau's officers.[75] Clinton and the new naval commander in North America, Admiral Thomas Graves, were persuaded to help Cornwallis only once Admiral Hood had arrived from the Caribbean with 10 more ships of the line. On 31 August Graves set sail for the Chesapeake.

De Grasse arrived before him, and disembarked his troops. On 5 September Graves and de Grasse fought a two-hour engagement off the entrance to the bay. Neither fleet lost a ship, but both sustained extensive damage. Rather than renew the battle, Graves decided to return to New York to refit. Even at this stage some British officers failed to grasp the seriousness of the situation. Captain Francis Reynolds of the Royal Navy was convinced that Cornwallis 'could hold out ... for ever, provided his Provisions will last'.[76] But the prospects for the besieged earl, as Graves himself fully appreciated, were now very bleak indeed. 'We cannot succour him, nor venture to keep the sea any longer', the admiral reported disconsolately.[77]

On land, Cornwallis was faced with a numerically superior enemy, even though Washington and Rochambeau had not yet arrived. On 2 September Clinton sent him warning of the allies' approach. At the same time he promised a diversion to draw Washington back to the north. This took the form of a particularly bloody descent on the unfortunate Connecticut coast, led by Arnold. The garrison of Fort Griswold, near Groton, was put to the sword, and most of New London was reduced to ashes. But Washington would not be deflected.

A desperate situation called for desperate measures. Clinton, casting caution aside, resolved that as soon as the fleet was again seaworthy, he would lead 5,000 troops to relieve Yorktown. In the past nervousness about the vulnerability of New York had deterred him from any bold strokes; but now he realized that he had to gamble everything to save

Cornwallis. He must, therefore, have been deeply frustrated when his rescue plans were stymied by the slowness of the naval repair work. The fleet did not re-embark – with over 7,000 troops on board – until 19 October. By then it was too late.

The allied army, around 16,000 strong, took post around Yorktown on the night of 28–9 September. Cornwallis abandoned his outer works to concentrate his forces. 'We now began to make preparations for laying close siege to the enemy', wrote one of Washington's sappers. 'We had holed him and nothing remained but to dig him out'.[78] The allied artillery started to bombard the British defences on 6 October. Johann Döhla, a German soldier in Cornwallis's army, recorded in his diary the awful experience of trying to withstand 'the terribly heavy cannonade'. 'Our command', he confessed, 'could hardly tolerate the enemy bombs, howitzer, and cannon balls any longer'.[79] On the night of the 14 October the French and Americans stormed two British redoubts. Cornwallis's position was now hopeless. A sortie to disable the besiegers' guns brought only limited and temporary respite. An attempt to break out via Gloucester, across the river from Yorktown, was wrecked by a storm. On 17 October – the very day four years before that Burgoyne had surrendered at Saratoga – Cornwallis proposed a negotiation of terms. Two days later the British marched out and laid down their arms. 'After all our efforts & Exertions the Garrison of York has at last been obliged to surrender', Lieutenant George Dunlop of the Seventy-fourth Foot wrote dolefully.[80] Even their enemies were moved by the pathos of the occasion: 'when I beheld them reduced from their former power & Consequence to their present miserable melancholy plight', Aedanus Burke of South Carolina confessed, 'I for a moment forgot their insolence, their depredations and cruelty'.[81]

*

'Charleston is our next object', exalted one of Washington's officers.[82] But over the following months Washington was unable to persuade the French to help him attack Charleston or New York. While the Americans could effectively deny the British access to the countryside around their remaining enclaves, they were in no position to do anything more adventurous without the co-operation of the French navy. And the French, although they did attack two of the Hudson's Bay Company's bases in August 1782, were unwilling to provide the necessary assistance because they wished to devote their energies to the war in the Caribbean.

A shortage of shipping for supplies and other logistical difficulties meant that the British were also unlikely to be able to launch another active campaign, even if they had been willing to do so. Many loyalists might continue to believe that America was recoverable, but others thought further military action was pointless. The familiar refrain that the chief beneficiaries of the war were the commissaries and quartermasters who were lining their pockets at the taxpayers' expense became more insistent. 'I think it high time to terminate the war in this Country', a depressed Frederick Smyth wrote after Yorktown, 'for in truth the war in this Country is now little more than a war upon the Treasury'.[83] Sir Guy Carleton, who replaced Clinton in May 1782, carried out his instructions to suspended offensive operations in North America. Savannah was evacuated the following July; Charleston in December. The British finally withdrew from New York on 25 November 1783.

Chapter 5 Notes

1. Lt. Col. Nesbitt Balfour to Lord Polwarth, 1 Feb. 1778, Polwarth Papers, Lucas Collection, L 30/12/3/4, Bedfordshire RO.
2. *DAR*, xv. 73.
3. For Carlisle's mission, see Chapter 9.
4. Journal of John Miller, entries of 5, 8 Dec. 1777, Joseph Reed Papers, New–York Historical Society.
5. Macleod to his aunts, 22 May 1778, Macleod Letters, Acc 6770, National Library of Scotland.
6. Laye Letters, 6807–154, p. 68, National Army Museum.
7. Brig.-Gen. James Pattison to Viscount Townshend, 11 May 1778, Mil. Doc. 7, p. 115, Royal Artillery Institution.
8. *The American Journal of Ambrose Serle*, ed. E. H. Tatum Jr. (San Marino, Ca., 1940), p.301.
9. *DAR*, xv. 126.
10. *Diary of a Common Soldier ... An Annotated Edition of the Military Journal of Jeremiah Greenman*, ed. Robert C. Bray and Paul E. Bushnell (DeKalb, Ill., 1978), p.122.
11. Alexander Hamilton to Elias Boudinot, 5 July 1778, *Hamilton Papers*, i. 512.
12. *Citizen Soldier: The Revolutionary War Journal of Joseph Bloomfield*, ed. Mark E. Lender and James Kirby Martin (Newark, N.J., 1982), pp.136–7.
13. For French thinking on this see Chapter 9.
14. Ens. Hugh Campbell to Duncan Campbell, 22 July 1778, Campbell of Barcaldine Muniments, GD 170/1118/8/1, SRO.
15. Sullivan to Henry Laurens, 1 Aug. 1778, *Letters and Papers of Major-General John Sullivan*, ed. Otis G. Hammond (3 vols., Concord, N.H., 1930–9), ii. 167.
16. 'The Wyllys Papers', Connecticut Historical Society, *Collections*, xxi (1924), 462.

17. 'Revolutionary Letters written to Colonel Timothy Pickering by George Williams of Salem', Essex Institute, *Historical Collections*, xliii (1907), 199.
18. Gambier to the Earl of Sandwich, 23 Oct. 1778, *Sandwich Papers*, ii. 324.
19. Capt. John Peebles to Dr Charles Fleeming, 26 Oct. 1778, Cuninghame of Thorntoun Papers, GD 21/495, SRO.
20. Murray to Elizabeth Smyth, 24 Oct. 1778, *Letters from America, 1773–1780*, ed. Eric Robson (Manchester, 1951), p.61.
21. Col. Charles O'Hara to Sir Charles Thompson, 20 Sept. 1778, Hotham Papers, DD HO 4/19, Hull University Library.
22. Erskine to Sir John Halkett, 21 Sept. 1778, Halkett Papers, MS 6410, fo. 64, National Library of Scotland.
23. 'Letters of Ebenezer Huntington, 1774–1781', *American Historical Review*, v (1899–1900), 720.
24. Clinton to Germain, 8 Oct. 1778, *DAR*, xv. 210.
25. Campbell's Journal, Campbell of Inverneill Papers, Canna House, Isle of Cana.
26. Campbell to William Eden, 19 Jan. 1779, Auckland Papers, Add. MS 34,416, fo. 246.
27. Maj. Henry Sheridan to [Vaughan], 24 July 1779, American Papers, Egerton MS 2,135, fo. 74, British Library.
28. Gadsden to Samuel Adams, 6 July 1779, *The Writings of Christopher Gadsden*, ed. Richard Walsh (Charleston, S.C., 1966), p.165.
29. *John Norton and Sons*, ed. F.N. Mason (Richmond, Va., 1937), p.430.
30. Collier to Clinton, [16 May 1779,] Clinton Papers, William L. Clements Library.
31. 'Letters of Joseph Clay', Georgia Historical Society, *Collections*, viii (1913), 141.
32. 'Meyronnet de Saint-Marc's Journal', trans. and ed. Roberta Leighton, *New-York Historical Society Quarterly*, xxxvi (1952), 279.
33. Francisco Collell to Gálvez, 26 Sept. 1779, *Spain in the Mississippi Valley ... Pt. I: The Revolutionary Period, 1765–81*, trans. and ed. Lawrence Kinnaird (Washington, D.C., 1949), p. 358.
34. Maj.-Gen. John Campbell to Germain, 15 Dec. 1779, *DAR*, xvii. 260.
35. *Journals of the Military Expedition of Major General John Sullivan against the Six Nations of Indians* (New York, 1970), p.88.
36. Tryon to Clinton, 28 May 1779, Clinton Papers.
37. Col. S.B. Webb to Joseph Barrell, 11 July 1779, *Correspondence and Journals of Samuel Blachley Webb*, ed. W.C. Ford (3 vols., New York, 1893–4), ii. 180.
38. *The Literary Diary of Ezra Stiles*, ed. Franklin B. Dexter (3 vols., New York, 1901), ii. 356.
39. *Revolution in America: Confidential Letters and Journals ... of Adjutant General Major Baurmeister*, trans. and ed. Bernhard A. Uhlendorf (New Brunswick, N.J., 1957), p.313.
40. Carleton Papers, PRO 30/55/17, 2088.
41. McLean to Clinton, 23 Aug. 1779, ibid., 2214.
42. Nooth to Lady Grimston, 23 Nov. 1779, Verulam MSS, D/EV F. 25, Hertfordshire RO.

43. Clinton to Cornwallis, 6 May 1780, Cornwallis Papers, PRO 30/11/2, fo. 28.
44. 'Papers of Gabriel Manigault, 1771–1784', ed. Maurice A. Crouse, *South Carolina Historical Magazine*, lxiv (1963), 11.
45. Dundas to Charles Dundas, 10 May 1780, in Margaret Isabella Dundas, *Dundas of Fingask* (Edinburgh, 1891), p.63.
46. Lt.-Col. Alexander MacDonald to his wife, 10 June 1780, Robertson-MacDonald Papers, MS 3945, fo. 61, National Library of Scotland.
47. 'Colonel Robert Gray's Observations', *South Carolina Historical and Genealogical Magazine*, xi (1910), 141.
48. Turnbull to Cornwallis, 15 June 1780, Cornwallis Papers, PRO 30/11/2, fo. 158.
49. Ibid., fo. 252.
50. Ibid., fo. 162.
51. *The Journal and Order Book of Captain Robert Kirkwood*, ed. J.B. Turner (Port Washington, N.Y., 1970), p.11.
52. *Hamilton Papers*, ii. 385.
53. Cornwallis to Germain, 19 Sept. 1780, *DAR*, xviii. 170.
54. 'Diary of Lieut. Anthony Allaire', in Lyman C. Draper, *King's Mountain and its Heroes* (Cincinnati, Ohio, 1881), p.511.
55. Capt. John Saunders to Lt.-Col. J.G. Simcoe, 14 Nov. 1780, Simcoe Papers, William L. Clements Library.
56. Letter-Books, Vol. I, Greene Papers, Library of Congress.
57. *American Campaigns*, i. 17.
58. 'Matthew's Narrative', *Historical Magazine*, i (1857), 105.
59. Rooke to Charles Rooke, 9 Oct. 1780, Rooke Papers, D 1833 F2/45, Gloucestershire RO.
60. *Sandwich Papers*, iii. 262.
61. 'Diary of Bernardo de Galvez', *Louisiana Historical Quarterly*, i (1917–18), 74.
62. Tarleton to his brother, 18 Dec. 1780, Tarleton Papers, 920 TAR 4/2, Liverpool RO.
63. Journal of Alexander Chesney, Add. MS 32,627, fo. 21.
64. O'Hara to the Duke of Grafton, 20 April 1781, Grafton Papers, Ac 423/191, West Suffolk RO.
65. Lt. Francis Dundas to Robert Dundas, 23 April 1781, H.M.C., *Report on the Laing Manuscripts* (2 vols., London, 1925), ii. 510.
66. *The Revolutionary Journal of Baron Ludwig von Closen*, trans. and ed. Evelyn M. Acomb (Chapel Hill, N.C., 1958), p.75.
67. *Jefferson Papers*, v. 361.
68. HMC, *Report on the Manuscripts of the Late Reginald Rawdon Hastings* (3 vols., London, 1904–34), iii. 193.
69. *The Revolution Remembered: Eyewitness Accounts of the War for Independence*, ed. John C. Dann (Chicago, 1980), pp.220–1.
70. 6 June 1781, Clinton Papers.
71. Gilbert to Lt. John Soule, 5 May 1781, *Winding Down: The Revolutionary War Letters of Lieutenant Benjamin Gilbert*, ed. John Shy (Ann Arbor, Mi., 1989), p.41.

72. Bannister to [Theodorick Bland], 16 May 1781, Bland Papers, Library of Congress.
73. 'The Journal of Ebenezer Wild', Massachusetts Historical Society, *Proceedings*, 2nd series, vi (1890–1), 144.
74. 'Journal of Lieut. William McDowell', *Pennsylvania Archives*, 2nd series, xv (1890), 301.
75. *Diary and Correspondence of Count Axel Fersen*, trans. and ed. Katherine Prescott Wormeley (London, 1902), p.48.
76. Reynolds to Lord Ducie, 28 Sept. 1781, Ducie and Morton Muniments, D 340a C28, Gloucestershire RO.
77. *Sandwich Papers*, iv. 183.
78. *Private Yankee Doodle*, ed. George F. Sheer (Boston, 1962), p.230.
79. *A Hessian Diary of the American Revolution*, trans. and ed. Bruce E. Burgoyne (Norman, Ok., 1990), pp.168, 172.
80. Dunlop to James Dunlop, 20 Oct. 1781, Dunlop Papers, box I, Library of Congress.
81. 'Correspondence of Hon. Arthur Middleton', *South Carolina Historical and Genealogical Magazine*, xxvi (1925), 187.
82. 'The Wyllys Papers', Connecticut Historical Society, *Collections*, xxi (1924), 472.
83. Smyth to Charles Blagden, 5 Dec. 1781, Blagden Papers S. 38, Royal Society.

6

The War Beyond America

In June 1781 Christopher Marshall, a Pennsylvanian supporter of the Revolution, wrote excitedly in his diary of the Spanish capture of Pensacola. He added, with more enthusiasm than accuracy, that in India 'Hyder Aly Kan' had retaken all the French posts which had been seized by the British.[1] Marshall was aware that the fortunes of his country were linked to events far from home. From 1778 the American war was not just an American war, but a world-wide struggle that stretched the British to the limit. The thinking of Lord North and his colleagues had to change dramatically. The American theatre – hitherto central to their concerns – now became just one of many, and by no means the first call on resources.

*

The West Indies absorbed 16 British infantry battalions between 1778 and 1783, besides the regiments under Major-General Grant taken from Clinton's command, while Clinton himself had to make do with a trickle of reinforcements. A considerable portion of the Royal Navy's strength was likewise committed to operations in the Caribbean.

Why was this so? The real value of the British islands to the home economy has been a matter of much debate among historians; R.B. Sheridan calculates that Jamaica alone was worth about £1.5 million a year to Britain on the eve of the war,[2] whereas Philip Coelho argues that the islands were more of a liability than an asset.[3] But most contemporaries had no doubt about the vital contribution of the West Indian colonies to British prosperity. The islands purchased significant quantities of British manufactures (many of which found their way to the Spanish and other foreign colonies in the region) and they sent home vast volumes of sugar, together with smaller cargoes of cotton, indigo and coffee. These imports freed the mother country from dependence on foreign products (90% of sugar imports were consumed within Britain itself) and provided the government with a buoyant customs revenue; it was small wonder that the security of the islands was a primary objective of British policy.

The French were equally certain of the value of their islands. If Jamaica – the biggest and most important of the British islands – was the brightest jewel in George III's crown, then St. Domingue was to the French the pearl of the Antilles. The West Indies probably accounted for about a third of all French external trade; they were responsible for the growing prosperity of the great western port of Bordeaux; and they added handsomely to Louis XVI's revenues. Indeed, given that the French islands produced their sugar more cheaply than the British, and French sugar therefore dominated European markets, the French perhaps had particular cause for alarm at the extension of the war to the Caribbean.

But both the French and the British, while having much to lose, had much to gain. For the French there was the promise of the return of the islands lost in the Seven Years' War and, with the capture of British islands, the undermining of the financial strength of their chief rival. For the British there was the chance to seize French islands, and with them more of the sugar trade, delivering at the same time heavy blows to French commerce and finances and perhaps knocking France out of the war. Hence the keenness of the British government early in 1778 to detach a significant portion of Clinton's army for an immediate attack on the French West Indies. And for the British there was also the opportunity, from 1779, of making gains at the expense of the Spanish, whose Central and South American empire was believed to be both rich and vulnerable.

The Caribbean was affected by the contest in America long before the French entered the fray. On 23 March 1776 an American squadron of eight small vessels under Commodore Esek Hopkins raided New Providence in the Bahamas and captured the British governor. And before 1778 the British were obliged to devote naval resources to trying to tackle the problem of American vessels collecting supplies from the West Indian islands. Dutch St. Eustatius was probably the key entrepôt, though other foreign islands also helped the Americans. Even the British sugar colonies were willing to trade with the rebels, for whom there was considerable sympathy amongst the British planters: a Royal Navy officer wrote from Barbados that half the people 'are of the strongest American principles'.[4] But the war in Caribbean began in earnest only with the outbreak of hostilities between the British and the French.

On 7 September 1778 the French attacked Dominica, formerly one of their islands, which had been in British hands since the Seven Years' War. With a garrison of less than 100 regular troops, and a white population that was pro-French, the British were quickly overwhelmed. In December

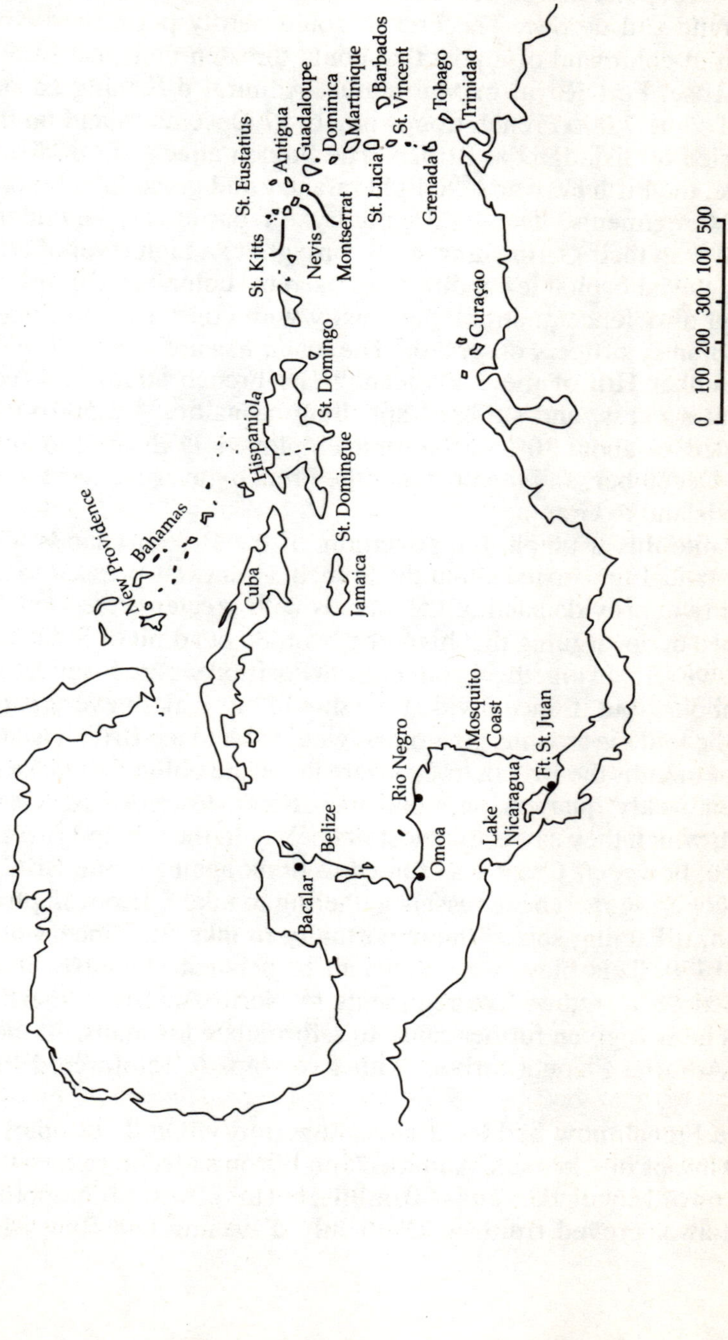

The war in the Caribbean and Central America

it was the turn of the British to go onto the offensive. The capture of St. Lucia, planned at the beginning of the year but delayed by Clinton's unwillingness to release the necessary forces, was finally accomplished. Grant's troops landed on 13 December and established a foothold north of the Grand Cul de Sac. The French could hardly permit the British to remain in command of a post that could threaten their main Caribbean naval base, Fort Royal in Martinique. Admiral d'Estaing accordingly landed some 7,000 French troops on 16–17 December, and on the 18th they tried to dislodge the British. The French attack fell on the British reserve, the Fifth Foot and the light infantry and grenadier companies of Grant's regiments. The light infantry, in particular, bewildered the French with their skirmishing skills, brought to a high level of effectiveness in the inhospitable conditions of the rebel colonies: 'this unexpected method of defence ... struck the enemy with consternation', one of the light infantry officers observed.[5] The result has aptly been described as the 'Bunker Hill of the Caribbean'.[6] The French attacked three times without success, and suffered appalling casualties – 1,500 killed and wounded, or about 30% of the force employed. D'Estaing re-embarked on 29 December, and the next day the French governor surrendered the whole island to Grant.

Despite this triumph, the governors of the British Islands were still understandably worried about the French. Grant was pressed to disperse his army to provide each of the islands with greater protection. This he declined to do, arguing that his force 'would not admit of Subdivisions – While we remain together in our present Position we are formidable to our Neighbours, but if once divided we should be weak in every part'.[7] The navy, he told the anxious governors, would shield the British islands. For several months the British troops were therefore obliged to make the best of their sickly quarters on what one officer described as 'this horrid Island; which they say is the most unhealthy in the whole West Indies'.[8] In time, however, Grant was forced to think again. While British ships were covering merchant vessels gathering to take Caribbean produce to Britain, d'Estaing seized the opportunity to take St. Vincent on 16–17 June 1779. This blow was sufficient to persuade Grant to ignore his instructions to return five regiments to North America. Shortly afterwards he was given further cause to reformulate his plans. At the end of June Admiral Picquet arrived with five ships to reinforce d'Estaing's fleet.

The French now had local naval superiority. On 2 July they landed 2,000 troops on Grenada. Admiral John Byron sailed to relieve the small garrison of 125 regulars and 400 militia; but his efforts to compel a French withdrawal proved fruitless. On 6 July d'Estaing and Byron fought a sharp action off Grenada. D'Estaing, according to a French report, 'won a complete advantage' and obliged Byron's squadron 'to take flight'.[9] Retiring to St. Kitts to refit, Byron confessed that 'their Masts Rigging and Sails appeared less injured than ours'.[10] The British troops on Grenada surrendered the same day. Grant now abandoned his earlier strategy of concentration, and started to disperse his forces: 'circumstanced as we are at present', he told Germain on 17 July, 'the Islands must fall if they are not defended by strong Garrisons'.[11]

In the western Caribbean the outbreak of war with Spain increased the pressure on the British. Jamaica was vulnerable to attack – a 'descent on this Island may be daily expected',[12] one of the inhabitants wrote in August 1779 – and the scattered British posts on the Central American isthmus were in immediate danger. A British force was sent by General John Dalling, the governor of Jamaica, to support the settlement at Belize and attack the Spanish base at Bacalar. The British arrived too late: on 15 September a Spanish expedition from Bacalar descended on St. George's Cay. The Jamaica force turned its attention to the fortress of Omoa, in present-day Honduras. On 20 October a motley crew of British troops, sailors and marines, British settlers in Honduras and local Indians stormed this strongly fortified post, meeting remarkably little resistance from the garrison: 'so great was the consternation and Cowardice of the enemy', gloated William Dalrymple, commandant of the Jamaica Corps, 'that it seemd as if they had lost the power of their arms'.[13] But only a few weeks later sickness had taken such a toll that Dalrymple's men were obliged to relinquish their prize and return to Jamaica.

Germain was all in favour of attacks on the Spanish empire, and he even managed to find some British regiments to devote to the task. Governor Dalling, however, was impatient; the prospect of Spanish booty seems to have encouraged him to act hastily. Early in 1780, without waiting for the promised reinforcements from Britain, Dalling despatched another force to strike at the Spanish possessions on the mainland. The objective this time was the San Juan River, the control of which, it was confidently thought, would secure a passage to the Pacific, where a base could be established to sustain predatory raids on the Spanish settlements along the west coast of Central and South America. Dalling's expedition, led by Lieutenant-Colonel Stephen Kemble of the Royal Americans, captured San Juan Castle, 69 miles upriver and near the shore of Lake Nicaragua, on 29 April. But, as in the previous October, British success proved transitory. Illness among the troops and sailors prevented further progress; and when the bulk of the British force withdrew to the coast, the Spanish easily retook the castle from the small garrison that Kemble had left behind.

Further east, meanwhile, the French and British fleets (now commanded by de Guichen and Rodney respectively) fought a series of indecisive actions. On 17 April 1780 Rodney claimed to have 'Compleatly beat' the French off Martinique, but he was prevented from capitalizing on his advantage by 'the crippled Condition of several Ships', particularly his own flagship, the *Sandwich*, which for 24 hours 'was with Difficulty kept above water'.[14] On 15–16 and 19–20 May Rodney chased and caught the French again; but, as in April, he was unable to score a clear victory. The British suffered almost as much damage as the French, and Rodney retreated to Barbados to refit on 22 May.

A potential crisis arose in June when Rodney failed to intercept a Spanish fleet under Don Josef Solano carrying more than 10,000 troops to the Caribbean. The Spanish left Cadiz on 28 April and reached Guadeloupe unmolested, only for the ravages of disease on the Spanish ships to rule out any offensive operations on their part. With so many of his crewmen and soldiers unfit for service, Solano decided to sail directly for Havanna. Once it became clear that the Spanish were in no position to menace Jamaica, and that de Guichen had returned to Europe, Rodney sailed north to assist Clinton and Arbuthnot at New York.

On land General Vaughan, Grant's successor, was unable to try anything adventurous. In March 1780 he had intended to recapture St Vincent and Grenada; the arrival of de Guichen's ships from Europe forced Vaughan to abandon this design. Soon after he reported that sickness was steadily sapping the strength of his command. From St. Lucia one of his officers wrote on 15 July that 'The unhealthiness of this Island has almost destroyed the few troops that came from America, and the recruits of new Corps are fast following them.'[15] Between 1 June and 31 August 568 British troops died on St Lucia alone. This 'great mortality', Vaughan lamented on 13 September, 'must effectively check all hopes of active operations – should it continue the army will be altogether inadaequate to those of defence'.[16]

Rodney's return from America in December, and news of the outbreak of hostilities with the Dutch, reactivated the British. Rodney and Vaughan moved quickly to comply with orders from London to take St. Eustatius. The island – and vast quantities of merchandise – fell to the British on 3 February 1781. The Dutch governor, unaware that his country was at war, offered no resistance. The neighbouring Dutch islands of Saba, St. Martin and St. Bartholomew also capitulated. Rodney was in no doubt about the importance of the conquest of St. Eustatius in particular: 'I hope this island will never be returned to the Dutch, it has been more detrimental to England than all the forces of her enemies, and alone has contributed to the continuance of the American war.'[17] Indeed,

Rodney was so dazzled by the brilliance of his triumph (and so impressed with the prize goods estimated to be worth at least £2 million) that he stayed at St. Eustatius until the end of April. This proved to be a serious mistake. Unsupported by Rodney, Admiral Hood, with only 18 ships of the line at his disposal, could do little to stop de Grasse bringing a powerful fleet of 24 capital ships from France to Martinique: 'not having the least prospect' of beating the French, and 'knowing the consequences of my being beaten', Hood prudently withdrew.[18]

De Grasse's arrival meant that the French were well placed to take the initiative. They felt strong enough, in fact, to act against St. Lucia, the British occupation of which was a constant irritation to the French. De Grasse landed a body of troops under marquis de Bouillé on St. Lucia on 10 May. Once on shore, however, de Bouillé discovered that the British had constructed formidable defensive works. Judging the island too well fortified, and perhaps mindful of the bloody outcome of the last French attempt to dislodge the British in December 1778, he decided to re-embark. Thwarted at St. Lucia, the French moved to Tobago at the end of the month. Rodney, finally back from St. Eustatius, tried to intervene; but by the time he arrived the garrison had already surrendered (2 June 1781).

Attention now turned to North America, with de Grasse sailing to the Chesapeake to help snare Cornwallis. As if Yorktown itself were not bad enough, it seemed to set in train a series of disasters for the British in the Caribbean. On 26 November 1781 de Bouillé surprised the numerically-superior garrison at St. Eustatius, the humiliated British commander blaming his defeat on the negligence of two of his sentries ('Fifteen minutes warning would have enabled me to have given a proper Account of the Enemy').[19] Saba, St. Martin and St. Bartholomew were seized by the French immediately afterwards.

To add to British troubles, the triumphant de Grasse returned to the Caribbean in December. Twice he tried to atttack Barbados, only to be driven back by unfavourable winds. He fixed on St. Kitts as a consolation prize. On 11 January 1782 de Grasse landed some 6,000 troops under de Bouillé. The British garrison retreated to a fortified post on Brimstone Hill. Both sides settled down to a siege, while de Grasse overran neighbouring Nevis. Hood sailed to relieve St. Kitts and on 25 January he fought a brilliant action with de Grasse, successfully outmanoeuvring the French admiral. As a result of this tactical victory, Hood was able on 28 January to land 2,400 troops to sustain the British defenders. But this force was insuffícent to oblige the French to lift the siege. Confronted by about 5,000 of de Bouillé's troops, the British retreated. They re-embarked on the 29th. Hood was convinced, despite this setback, that Brimstone Hill could hold out. He was over optimistic. After a month's siege, and with scarcely 500 men fit for duty, the British garrison surrendered on 13 February. De Grasse then compelled the surrender of Montserrat before sailing back to Martinique. As if to confirm the low level to which British fortunes had sunk, in March the Spanish took the British post on the Rio Negro in modern-day Honduras.

The French now began preparations with the Spanish to attack Jamaica. But the Bourbon position of superiority was about to be undermined. The balance of power had already shifted. Thanks to the timely arrival of further ships from Britain, by 25 February the combined fleets of Rodney and Hood outnumbered de Grasse's vessels. De Grasse left Fort Royal on 8 April with a great convoy of transports carrying 10,000 French troops for the Jamaica expedition. Hood and Rodney slipped anchor from St. Lucia and set off in pursuit. After two days of manoeuvre, de Grasse was compelled to offer battle on 12 April near the Isles des Saintes, in the channel between Dominica and Guadeloupe. The action proved to be one of the most decisive naval engagements of the eighteenth-century. The two sides exchanged fire for nearly an hour and a half, and then Rodney 'forced his way through the Enemys Line'.[20] When the French line was broken again, their centre was subjected to a concerted attack. The captain of a merchantman from Ostend, which was in the vicinity at the time, reported to the French authorities that the battle must have been 'very bloody'. That night he saw a vessel burning 'which in fact exploded totally about an hour later'.[21] He witnessed only part of the French disaster. By nightfall de Grasse's flagship, the *Ville de Paris*, had surrendered. Four other French ships were taken, plus all the heavy artillery intended for the attack on Jamaica. Hood pressed for the British to continue the pursuit, and complained bitterly when Rodney refused ('The very important and favourable opportunity that has been lost ... is melancholy in the extreme to think of').[22] French sources seem to support Hood's judgement: Carl Gustaf Tornquist, a Swede serving with de Grasse, was later of the opinion that had Rodney 'followed Admiral Hood's injunction to pursue throughout the whole night, the victory of the English would have been still greater'.[23] Even so, Rodney had won a notable triumph. 'You may now despise all your enemies', he told the Earl of Sandwich. The French fleet had been given 'such a blow as they will not recover'.[24] The projected descent on Jamaica was abandoned; and Rodney sailed into Port Royal on 28 April to a hero's welcome.

Decisive as it was, the battle of the Saintes did not end operations in the Caribbean. When the attack on Jamaica was shelved, the Spanish forces assembled on Hispaniola re-directed their attention to the Bahamas. On 8 May New Providence, with a garrison of a mere 170 men, surrendered to a Spanish invading force of 5,000 troops. At Jamaica, Archibald

Campbell, the new governor, was sure that the French and Spanish would renew their efforts to take his island once the hurricane season ended in November. The London government, too, believed that the West Indian conflict, far from winding down, was set to become more important than ever. General Carleton, Clinton's replacement as commander-in-chief in America, was ordered in August 1782 to send all the troops he could spare to the Caribbean, and to take command of them himself. From off Monte Christi, on the north coast of Hispaniola, Hood was delighted to write that reports of these orders had 'raised fears not only in the Spaniards but in the French also'.[25] But Hood was to be disappointed. Lack of shipping prevented Carleton from responding to his instructions, and it was in fact the French troops in North America who sailed south in December 1782. They reached the Caribbean in February 1783. But any chance of realizing the long-standing ambition of taking Jamaica was scotched when news arrived of the signing of the peace preliminaries. The last action of the war in the Caribbean was the recapture in April 1783 of New Providence by a contingent of American loyalists led by Colonel Andrew Devereaux.

*

Europe was not a major theatre of operations during the American war, but there were notable naval battles off the Iberian coast, in the Channel and North Sea; Gibraltar and Minorca were besieged and Jersey attacked; and Britain itself was threatened with invasion. Between 1778 and 1783 the British had no choice but to use precious resources for home defence and the protection of their Mediterranean posts: a considerable Channel fleet was kept in being, and behind it a good portion of the regular army waited to counter an enemy landing – indeed, about half of Britain's military manpower (if one includes the embodied militia) was allocated to this duty.

Even before France's official entry into the war, the potential danger of a sudden French descent on England ensured that sizeable naval and military forces had to be kept at home. In the months preceding open hostilities, the Earl of Sandwich, the first lord of the admiralty, became more and more convinced that an invasion attempt was imminent. When d'Estaing sailed for America with the Toulon fleet in April 1778, Sandwich believed that a juction between d'Estaing and the main French Atlantic fleet at Brest was in the offing. Sandwich was so worried that this junction would threaten British command of the Channel that he delayed sending out Byron to reinforce Lord Howe until there was no doubt that d'Estaing was heading west rather than north.[26] Nor should we forget that the fledgling Continental Navy and American privateers took the fight into British waters. In April 1778 John Paul Jones raided Whitehaven, tried unsuccessfully to abduct the Earl of Selkirk from his Scottish home (he was away at the time), and finally captured the Royal Navy sloop *Drake* off Belfast Lough. Jones continued his exploits the next year – capturing merchant prizes in the North Sea; threatening to destroy Leith; and on 23 September 1779, after a long and sanguinary engagement off Flamborough Head, taking H.M.S. *Serapis*, a 50-gun frigate.

By this time, however, the British had more formidable foes with which to contend. The combined French and Spanish fleets easily outnumbered even the significant quantity of British ships that the admiralty had carefully husbanded in home waters. The Royal Navy's weak position was not so evident in 1778, when Britain faced only France; all the same, it was with difficulty that the British retained control of the Channel. On 13 July the Brest fleet under Admiral d'Orvilliers appeared briefly off the Isle of Wight, causing considerable alarm, before moving down the Channel to engage the ships commanded by Admiral Keppel. After several days of manoeuvering for position, the two fleets met off Ushant on 27 July. Keppel outnumbered d'Orvilliers by 30 ships of the line to 27; but the British were unable to prevail in a gruelling battle. The French, Keppel reported, adopted the tactic of 'disabling the Kings Ships in their Masts & Sails'.[27] They were largely successful: one of Keppel's junior officers, Lieutenant Edward Bowater of the *Terrible*, wrote of how the fire from the French had 'Shatter'd our Masts Rigging and sails all to pieces'.[28] The two sides broke off the action at nightfall, and Keppel's demobilized ships were in no condition to follow d'Orvilliers. Keppel expected the French to renew the battle the next day. Their failure to do so left the British as nominal winners. In truth, the two fleets had merely battered each other into a state of mutual exhaustion.

It was widely believed in Britain that an invasion had been averted; but the real threat came the following year. In 1778 the French massed troops along their coast in order to persuade the British not to send naval reinforcements to America. In 1779 the assembly of an army in Brittany and Normandy was no feint; the French intended to send their troops across the Channel. They had been planning such an attack for some time. Vergennes thought it unnecessary – indeed, undesirable – for Louis XVI's army occupy London. A landing in Ireland, where the Catholic majority could be expected to be well-disposed, was a possiblity; preferably, however, he wanted a destructive raid on a British naval base: Portsmouth was favourite, with Plymouth an acceptable alternative. Such a raid, the French believed, would have dramatic consequences. British public opinion would be shaken, and the system of credit that sustained the British war effort would therefore be undermined. It was the Spanish entry into the war that pushed these plans up the French agenda, for, as a condition of committing their country to the struggle, the Spanish government pressed for a joint assault on England with the aim of forcing the British to come to terms.

The Royal Navy missed the opportunity to prevent a junction of the French and Spanish fleets when Admiral Sir Charles Hardy failed to intercept d'Orvilliers's ships leaving Brest. The French admiral was near Corunna, his rendezvous point, by 10 June 1779. At Le Havre and St. Malo more than 30,000 French troops were waiting to be transported across the Channel. D'Orvilliers was obliged to wait for six weeks for the Spanish to join him from Cadiz, and the combined fleet – with some 66 ships of the line – did not sail north until 30 July. But by 16 August the French and Spanish were in sight of Plymouth, with Hardy – commanding only 39 ships – far away in the west trying to block any attempted descent on Ireland. The moment of truth appeared to have arrived: 'God preserve Old England', an apprehensive Berkshire squire wrote in his diary.[29]

Hardy was instructed to abandon his defensive position and, if circumstances seemed favourable, to attack. But strong winds drove him farther west, allowing the Armada to lie unmolested off Plymouth. On shore, feverish preparations were made to repel a landing. On 24 August Commissioner Ourry reported from Plymouth dockyard that he had received help from more than 600 Cornish tin miners, who were busy 'deepening the ditch without the lines and putting them in a proper state of defence'. Arms had been distributed to local volunteers, and 'numbers of useful men are flocking in'.[30] An anxious local resident explained in a letter composed near Tavistock on 29 August that he was preparing to send his wife, daughter and father to safety and, in the event of a landing, to go with his sons to 'join the troops & stand or fall with them'. With a bitterness born of apprehension, he wrote: 'We all consider ourselves deserted by Hardy's fleet'.[31]

Hardy was at last able to sail back into the Channel; but, rather than engaging the combined fleet, he chose to retire to Spithead on 3 September. The humiliation was too much for Sir William Meredith, an opposition MP waiting anxiously at Portsmouth. 'English shyness is a revolution in the maritime System of Europe', he wrote on 13 September.[32] The king was equally dismayed; the Franco-Spanish fleet had to be dealt with severely, he told his ministers, if British pride were to be restored. North and some of his colleagues were less sure; a victory would be splendid, of course, but the risk was too great. Their argument, though born of timidity, did not lack strength. There could be no invasion while there was still a British naval force able to attack the transports ferrying the troops from Brittany and Normandy. Only when d'Orvilliers had decisively defeated Hardy could he safely bring the waiting army across the Channel.

Unheroic and unpopular as it was, the strategy of patient avoidance of battle paid off. Relations between the French and Spanish were growing worse by the day, not least because the French could not refrain from contemptuous criticism of the poor seamanship of their allies. Worse still, several of the allied ships had been damaged in the voyage through the tempestuous Bay of Biscay and were barely seaworthy. Many of the sailors were desperately sick. By the middle of September, debilitated by disease, the French and Spanish crews were too weak to stay at sea any longer. D'Orvilliers had no option but to sail for home. After weeks of high tension, British fears of invasion could now recede.

Elsewhere, however, the pressure was maintained. Spanish involvement in the struggle put the exposed British garrisons at Gibraltar and Minorca at great risk. The Spanish had begun a blockade of Gibraltar as soon as they entered the war. Gibraltar, in particular, was a strong post, unlikely to fall quickly; but to sustain the Mediterranean outposts it was imperative to send out reinforcements and supplies. By the end of August 1779 the governor of Gibraltar, George Augustus Eliott, was urgently reminding the secretary of state that 'very little fresh Provision is left'.[33] Admiral Rodney, after having escorted a West Indian convoy 400 miles west of the Bay of Biscay, sailed south on 4 January 1780 with succour for Gibraltar and Minorca. Four days later he captured a Spanish provision convoy *en route* from San Sebastian to Cadiz. On 16 January he engaged with a Spanish squadron under Don Juan de Lángara. The outnumbered Spanish were convincingly beaten: 'one of the Enemy's Line of Battle Ships Blew up with a Dreadful Explosion', Rodney told the secretary to the admiralty board;[34] and after a pursuit which continued until the early hours of the next morning a total of six Spanish vessels were captured. Rodney then drove off the blockading squadron and safely landed the troops and provisions at Gibraltar.

Rodney's successful voyage made him as much of a hero in Britain as he was to be in Jamaica two years later. But vital as his arrival was to the beleaguered garrison of Gibraltar, it brought only temporary respite. The Spanish stepped up their siege in late November 1780, carrying their trenches forward against the British positions on the Rock. Food supplies again ran short, as the Spanish attempted to stop provisions reaching the British from North Africa. Another relief convoy was clearly required. Just in time, Admiral Darby brought in nearly 100 supply ships in April 1781. No sooner had he done so than the Spanish began a heavy bombardment, which they kept up with scarcely an intermission. A recent arrival at this time wrote of the 'pitiful situation' of the town. 'The Spanish batteries have fired without slackening', and nearly every house was destroyed.[35] This added considerably to the difficulties facing Eliott. Soldiers and civilians were now forced to camp out on the rock. Spanish shelling, furthermore, had exposed cellars full of wine, and Eliott's troops engaged in an orgy of drunken looting: 'Robbery, Moroding, Housbreaking and Shopliftling by the Soldiery', a British diarist recorded.[36] But despite this breakdown of discipline, Eliott was confident that his men would continue to resist the Spanish: 'I have no reason to suspect that any care is wanting at the Posts next the Enemy; the Soldier is wakefull when the fumes of liquor are evapourated'.[37]

If Gibraltar and Minorca seemed vulnerable, the Channel Islands were perhaps still more so. With none of the natural defensive strength of Gibraltar, and only a few miles from the Normandy coast, they seemed almost to invite attack. The moment the Anglo-French war started, fears were expressed for the safety of the islands, and ships had to be made available for their protection: 'the idea of an Invasion of Jersey & Guernsey is so well founded', Sandwich wrote to Captain Francis Reynolds of the Royal Navy on 28 August 1778, '& their defence of such public importance, that I cannot answer sending you elsewhere'.[38] It was not until May 1779, however, that the anticipated attack took place. This was easily repulsed but in January 1781 a French invasion force of 2,000 troops tried to take Jersey. Stormy weather meant that only 800 men could be landed. The British were unprepared, and the French seized St. Helier. Had it not been for the inspired leadership of young Francis Peirson, the major of the newly-raised Ninety-fifth Foot, the British might not have rallied. After fierce fighting in St. Helier, in which Peirson lost his life, the invaders were expelled.

A further diversion was caused by the Dutch. Since the British declaration of war in December 1780 the fleet of the United Provinces had remained at the Texel, leaving Dutch merchant shipping at the mercy of the Royal Navy and British privateers. In response to an intense domestic clamour, and perhaps emboldened by the recent capture by a Dutch captain of a British ship off Gibraltar, on 1 August 1781 Admiral Zoutman eventually put to sea with eight men-of-war and supporting frigates to convoy 72 merchantmen to the Baltic. Four days later Zoutman made contact at Dogger Bank with a British squadron of nine ships of the line under Admiral Hyde Parker. The two sides slugged it out for three and a half hours, sustaining heavy casualties. Like punch-drunk pugilists, they disengaged and returned, bruised and bloody, to their respective bases. Both the British and the Dutch claimed a victory – ' the honour of our flag was not sullied, but rather to the contrary. The Fatherland's honour ... was maintained and assured', the Prince of Orange was informed[39] – but the balance of advantage was with the British. The Baltic convoy had been forced to return to the Texel, and the Dutch fleet did not engage again with the Royal Navy.

A more formidable naval threat appeared shortly afterwards when another Franco-Spanish armada occupied the Channel approaches. The ships of the combined fleet were able effectively to block the western entrance to the Channel by forming a line stretching from Ushant to the Scilly Isles. Sickness on board the allied vessels, together with the caution of the aged Spanish Admiral Don Luis de Córdoba, prevented the French and Spanish making the most of their position; but Sandwich was forcefully reminded – not that he had forgotten – of the vulnerability of the south coast and the need to keep strong forces in home waters.

The Franco-Spanish fleet that so alarmed the British admiralty had sailed from the Mediterranean, after disembarking soldiers and equipment on Minorca. On 19 August 8,000 Spanish troops were put ashore and the next day they began a formal siege of Fort St. Philip at Port Mahon. They were joined by a similar number of French soldiers brought from Toulon. Just before the siege began, Lieutenant Archibald Cuninghame of the Fifty-first Foot had painted an unfavourable picture of the condition of the regiments based in Minorca: most of the officers, he told his sister, 'border upon antiquity', while the men of his corps 'are entirely worn Out. I dont suppose were we in England now that above fifty men of the whole Reg^{t} would be kept.'[40] Yet the British and Hanoverian garrison was in no immediate danger, for food stocks were high and the defences were strong. In September the defenders even made a bold attack on the investing army, destroying a number of batteries and carrying off 100 prisoners. General James Murray, the British governor, confidently boasted that the duc de Crillon, commander of the allied forces, 'will find Fort St. Philip's a harder nut to break than he imagines'.[41] The besiegers steadily built up their strength; by the end of December one of de Crillon's senior officers, baron de Falkenhayn, announced colourfully that 'In a few days we will be able to salute Fort St. Philippe with more than 100 mouths of fire'.[42] But Murray's position was weakened by scurvy rather than the efforts of the Spanish and the French. By 1 February 1782 only 760 British and Hanoverian troops were fit for duty. Over the next two days another 100 were hospitalized. Murray now made one last effort to persuade the besiegers to decamp. His artillery fired incessantly on the allied trenches throughout the night of 3–4 February. When it became obvious that this desperate ploy had achieved nothing, Murray agreed to discuss terms. After such a staunch defence,

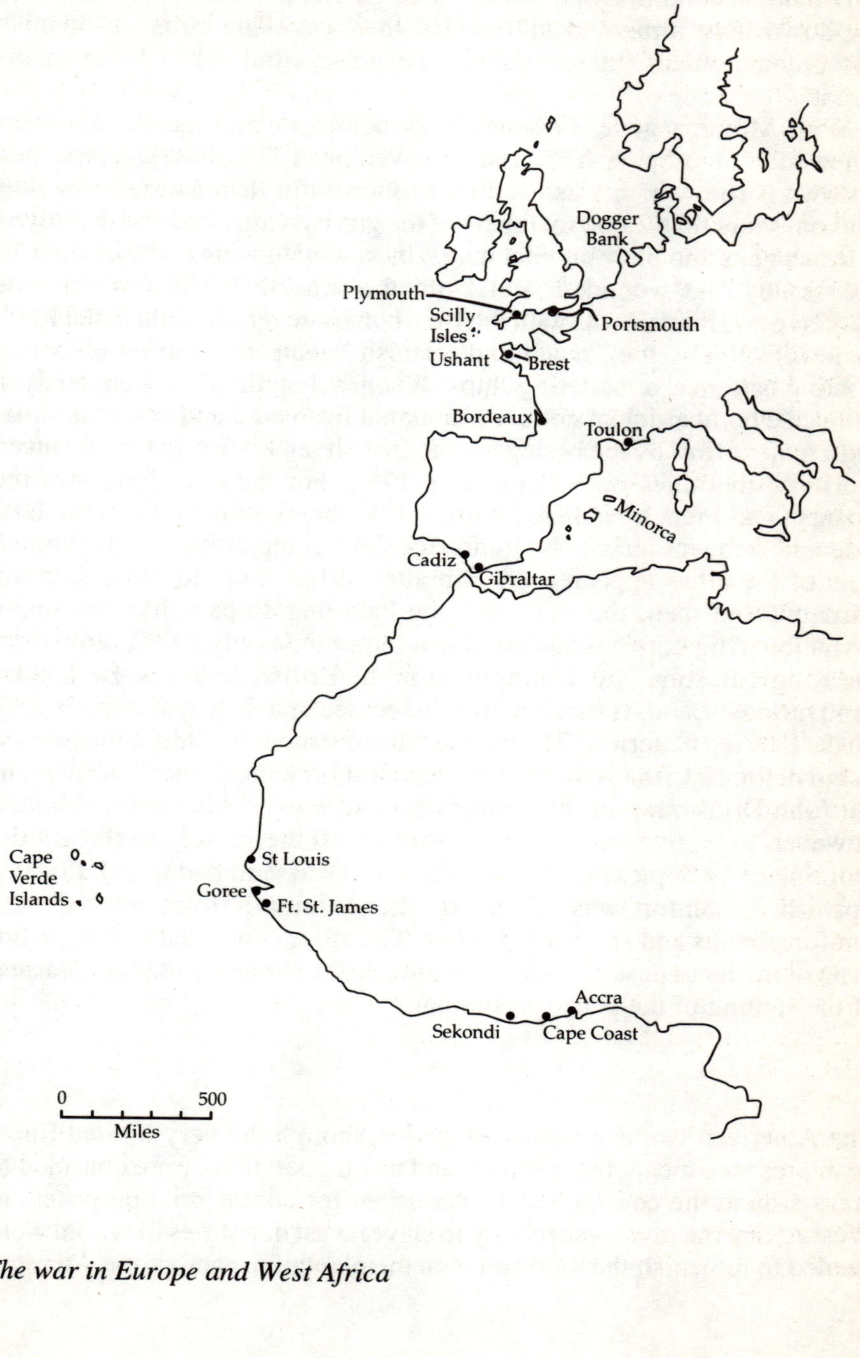

The war in Europe and West Africa

the French and Spaniards were inclined to be generous. The sickly garrison was permitted to march out with their flags flying, and to return to Britain on condition that they took no further part in the war. 'As they lay down their arms and unbuttoned their cartridge belts', a Spanish observer recorded, 'there was no one present but felt a lump in his throat'.[43]

With Minorca gone, Gibraltar was bound to become the focus of renewed Franco-Spanish efforts. In November 1781 Eliott had launched his own sortie, which was even more successful than Murray's at Port Mahon. More than 2,000 members of the garrison attacked and destroyed Spanish guns and blew up vital magazines, while losing no more than 30 men killed and wounded. But from the time that Minorca fell, the initiative at Gibraltar was with the Bourbon besiegers. To attack the Rock more effectively, the French and Spanish began to construct elaborate floating batteries, or battering ships. When, at length, these were ready, a full-scale bombardment could be mounted from land and sea. Anticipating a major effort by the besiegers, the British rained fire on the advanced works of the allies on 8 September 1782. For the next few days the Spanish and French retaliated with a heavy bombardment from the land side and frequent attacks by gunboats. On 12 September the combined fleet of the allies appeared off Gibraltar in full array to intimidate the garrison; and then, the next day, the battering ships – like 'so many invincible Aligators'[44] – anchored in a semi-circle only 1,000 yards from the King's Bastion, the central point in the British defences. Both sides fired furiously, and, at first, the British seemed unable to make any impact on the floating batteries. 'The wonderful construction of the ships seemed to bid defiance to the powers of the heaviest ordnance', recalled Lieutenant John Drinkwater of the Seventy-second Foot.[45] After several hours, however, extensive use of red-hot shot set all the vessels alight. By the morning of 14 September it was clear that the assault had failed. Franco-Spanish discomfort was increased when Admiral Howe arrived with reinforcements and stores in October. The allies continued the siege, but in no more than a desultory manner, until on 2 February 1783 they learned of the signing of the peace preliminaries.

*

The American war even affected Africa, though the very limited European presence meant that military and naval operations were confined to areas around the coastal bases established for commercial purposes. In West Africa the trade was mainly in slaves, vast quantities of whom were needed to replenish the workforce on the plantations across the Atlantic.

The war along the West African coast, then, was essentially about securing or improving access to supplies of slaves, and disrupting or even cutting off the supplies of one's enemies.

The peace of 1763 had given the British the French colony of Senegal in West Africa. This was added to the neighbouring British post at James Island to form the new royal province of Senegambia. The French had retained Goree, but with the renewal of hostilities they moved to recover Senegal. In January 1779 an expeditionary force surprised the weak British garrison at St. Louis, and the next month the French compelled the surrender of James Island and destroyed the British fort. Later the same year, however, Admiral Sir Edward Hughes, *en route* to India, landed marines from his squadron to occupy the fort at Goree, which the French had abandoned. Thereafter the two sides remained in possession of the posts they had taken.

Farther south, on the Gold Coast, a French frigate in May 1779 raided and severely damaged the British base at Sekondi. This region became the scene of more activity once the Dutch war broke out. Early in 1781 the British government raised two independent companies of troops, largely from convicts, with a view to sending them out to protect the British forts and, if possible, to seize the more numerous Dutch posts. Captain Kenneth Mackenzie, the commander of the force, stopped off at Goree and collected reinforcements from the garrison, and arrived at Cape Coast Castle on 5 February 1782. After consultation with the governor, Mackenzie resolved to make an attempt on the Dutch forts at Elmina and St Jago. The attack, carried out later that month, was a disastrous failure, which, according to the report of the senior naval officer present, revealed Mackenzie's unsuitability for command. But in March the British were more successful, securing the surrender, without any resistance, of Mouri, Kormantin, Apam and Beraku. The Dutch garrison at Accra offered determined opposition – the British lost nearly 150 killed and injured in the siege – but on the night of 18–19 March the Dutch abandoned the fort. In May the Dutch surrendered their post at Kommenda.

A much more important prize, far to the south, eluded the British. The colony of the Dutch East India Company *(Vereenigde Ooste-Indische Compagnie)* at the Cape of Good Hope sat astride the sea route to India and had been selected as a target by the London government as soon as the Dutch war began. Indeed, so important was the Cape that plans drawn up in 1780 for taking Spain's Pacific territories were shelved, and the naval and military forces assembled for this purpose were diverted to the southern African expedition. Commodore George Johnstone sailed from Britain with a squadron of ships and some 3,000 troops in March 1781.

The French, however, were aware of British intentions, and at the same time they despatched a naval and military force under Admiral de Suffren. On 16 April de Suffren came across Johnstone's squadron, taking in water and undergoing repairs, at Porto Praia in the Cape Verde Islands. De Suffren decided to enter the harbour to attack the British ships. After a fierce engagement the French withdrew. They had done enough damage to delay Johnstone's departure, and to this extent de Suffren's daring tactics paid off. 'I have a thousand apprehensions', one of Johnstone's captains wrote on 20 April, 'that Monsieur Suffren will (however Disabled) reach the cape before us. If so, farewell to our Expedition and all our Golden dreams.'[46] De Suffren did indeed beat the British to the Cape, arriving there on 23 June. He immediately landed two French regiments to reinforce the Dutch garrison. Discovering that the Cape was now strongly defended, the British abandoned their project. Johnstone returned to St. Helena and from there sailed home, while General Medows took the troops on to India.

*

India, and the Asian trade associated with it, was not as valuable to the British as their West Indian islands; but it was still very much worth defending. In 1772 Asian exports to Britain amounted to £2.4 million and made a far from negligible contribution to government revenues. Although the East India Company had substantial armed forces of its own, eight British regular regiments and two Hanoverian battalions arrived as reinforcements between 1779 and 1783, together with naval squadrons under Admiral Hughes and Sir Richard Bickerton.

The French likewise had cause to defend their trade, which had expanded rapidly since the demise of the monopolistic *Compagnie des Indes* in 1769. A French naval presence in the Indian Ocean, moreover, could help to disrupt Britain's trade with Asia, and therefore her prosperity and war finances. If this was what the French intended, they were at least partly successful. In 1781 Britain's Asian imports were higher than in any of the five years preceeding the war; but in 1779 and 1780 they fell below £1 million, and in 1782 to only £626,000. And the French, as we have seen, had other considerations in mind. War in India tied down British resources that might have been profitably employed elsewhere – in America, in the Caribbean, in Europe, or against the Dutch possessions in the East Indies.

The American war came to India when news arrived of the outbreak of hostilities between Britain and France. The British East India Company's forces at Bombay were already embroiled in conflict with the powerful

Maratha confederacy of Indian princes, a conflict that was to drag on until May 1782 and drew in a good portion of the company's main Bengal army. Even so, the company was determined to take advantage of the Anglo-French war, and most of the French posts were rapidly occupied. Pondicherry, the most powerful, was reduced only after an 'obstinate Defence' lasting 77 days.[47] By the end of October 1778, Mahé, on the western (or Malibar) coast, was the sole remaining French factory in India, and in March 1779 it fell to an expedition from the British base at Madras led by Lieutenant-Colonel John Braithwaite. But in their anxiety to eliminate their European competitors, the British provoked a war with another Indian ruler, and that war in turn provided the French with an opportunity to strike back at the British. Haidar Ali of Mysore was a long-standing rival of the Nawab of the Carnatic: both princes were ambitious to extend their influence in southern India. As the Nawab had allied himself to the representatives of the British East India Company at Madras, Haidar – to counter the Nawab and the British themselves – adopted a pro-French stance. He regarded the post at Mahé as under his protection and was understandably offended by its seizure. In retaliation he decided to attack the territories of the Nawab.

In July 1780 Haidar's enormous army poured into the Carnatic. His cavalry proceeded to plunder and lay waste the countryside and seized many of the towns. Sir Hector Munro, the Madras army commander, belatedly left St. Thomas Mount, near Madras, on 26 August to relieve the garrison at Arcot. Haider Ali raised the siege to concentrate his forces against Munro's advance. A British force under Lieutenant-Colonel William Baillie, marching from Guntor Circar to the north to join Munro, was attacked by Haidar's troops on 6 September. Baillie's detachment beat back their assailants, but Baillie was prevented from reaching Munro by the state of the local rivers, which had been swollen by heavy rain. Munro despatched a party of grenadiers and light infantrymen to sustain Baillie and then on 10 September, hearing firing in the distance, the whole of Munro's army set off in the direction of Baillie's beleaguered force. As Munro's men approached, they received the chilling news that Baillie had been totally defeated and obliged to surrender. It seems that while his men were fending off the Mysorians, Baillie's ammunition carts exploded. Taking advantage of the confusion, Haidar's cavalry inflicted a terrible slaughter, which, according to some accounts, ceased only on the insistence of his European advisors. Nearly all of Baillie's 4,000 troops were killed or captured. Munro's army, to which Haidar's victorious cavalry now turned its attention, began a hasty withdrawal: 'never was so Scandalous so Dishonourable a retreat made by British Troops', the shaken Major James Mackenzie of the Seventy-third Foot reported.[48]

Abandoning their baggage, the British arrived back at St. Thomas Mount on 14 September. As a young ensign in Munro's force gloomily reflected, 'The loss of Col: Baillies Army is the severest blow the English ever sustained in India ... my head is so filled with it that I can think of nothing else'.[49]

The precariousness of the British position at Madras was recognized farther afield. At Calcutta a British medical officer commented that 'There appears to be nothing wanting but an European enemy in concert with the country powers, to hurl destruction among the Company's possessions in that part of the world.'[50] In fact the French governor of Isle de France in the Indian Ocean had already opened negotiations with Haidar Ali. The French had sent out ships and troops to reinforce the garrison of Isle de France and Isle de Bourbon in December 1778, and now they could have intervened decisively. Yet they let the opportunity pass. A French squadron, commanded by Admiral d'Orves and with troops on board, appeared off the Coromandel coast at Cuddalore on 25 January 1781. Haidar asked d'Orves to land the soldiers and to keep his ships at hand to co-operate against the British; but d'Orves, unwilling to commit himself so far from his base, decided to return to Isle de France to collect a more formidable force.

By this time, Sir Eyre Coote had arrived at Madras with reinforcements from the Bengal army. He started offensive operations in the new year, and on 1 July 1781 Coote scored a notable victory over Haidar's forces at Porto Novo south of Cuddalore. This proved to be the turning point for the British. Coote followed up his success at Porto Novo with a more 'dearly gain'd' triumph at Polilur on 27 August, on the very site of Baillie's defeat the previous September ('a discouraging Consideration' for the Company's troops when the action began, one of Coote's officers confessed).[51] Coote completed the sequence by scattering Haider's army at Sholinghur on 27 September. But, important as the psychological impact of these victories undoubtedly was, they failed to push Haidar out of the Carnatic. Lack of cavalry, and perhaps still more seriously shortage of money, provisions and transport, made it difficult for Coote to follow up his successes: 'we are more like beaten than victorious Troops', he complained after the battle of Porto Novo.[52]

In 1781 news reached India of the outbreak of the Anglo-Dutch war. Just as in 1778, the British moved quickly to seize the trading posts of their rivals. In June and July all the Dutch factories on the mainland were taken, with the exception of the powerful fortress of Negapatam – the Dutch equivalent of Pondicherry. Negapatam was blockaded by sea by Admiral Hughes from 8 July, and on 30 October a formal siege began from the land side by East India Company forces under Munro. On 10

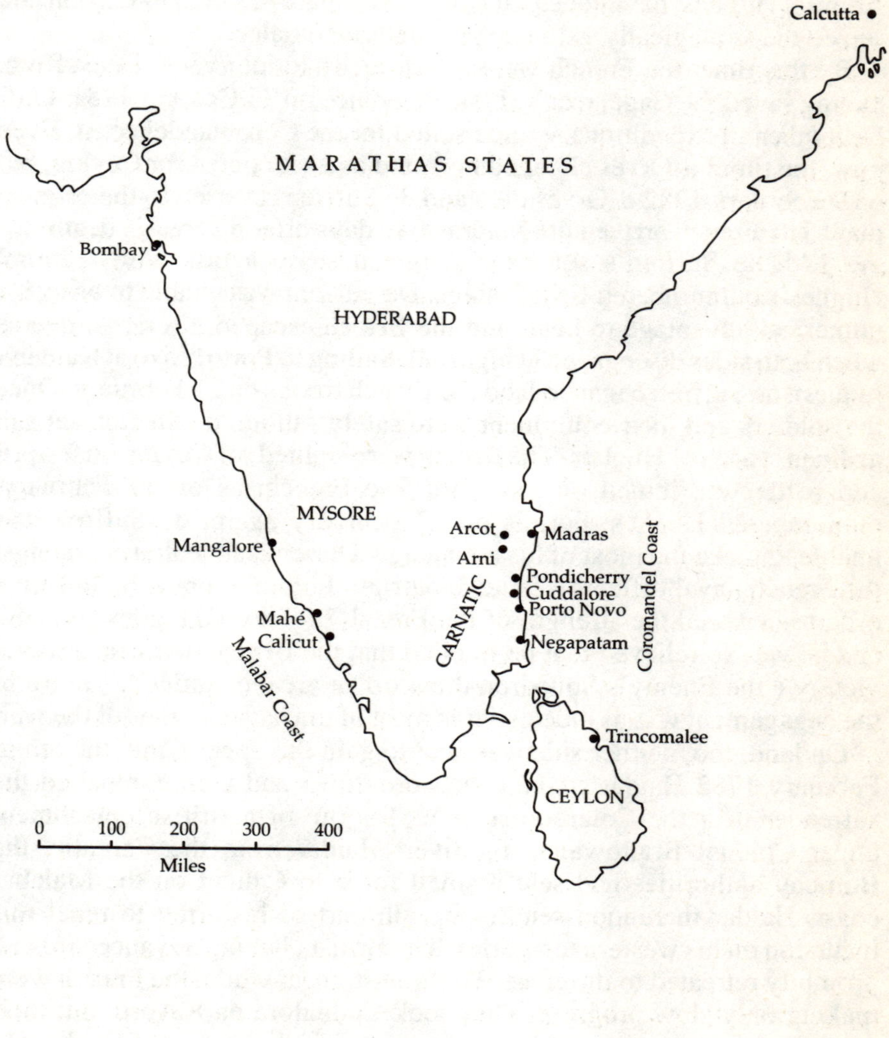

The war in India

November the defences were breached and, after a fruitless sortie 're-pulsed with great Loss on their Part & some on ours',[53] negotiations were opened on the 11th. The garrison – some 6,500 strong, if we include the troops lent by Haidar Ali – capitulated the next day. Beyond India itself, the British were also successful in taking the bases of the *VOC*. In August 1781 East India Company troops and ships captured Padang and with it all the forts and factories on the west coast of Sumatra; and in January 1782 Admiral Hughes, having helped to take Negapatam, sailed to Ceylon and seized the strategically valuable port of Trincomalee.

By this time, the French were at last ready to intervene. De Suffren, having saved the Cape, reached Isle de France on 25 October 1781. On 7 December an expeditionary force sailed for the Coromandel coast. Even now, the timid d'Orves changed his mind and set course for Ceylon; but on 9 February 1782 d'Orves died and de Suffren reverted to the original plan. The French arrived off Madras five days after d'Orves's death. On the 17th de Suffren's squadron fought a sharp action with Admiral Hughes's outnumbered British ships. De Suffren was unable to bring his numerical advantage to bear, and the British escaped a serious reverse when both sides disengaged at nightfall. Sailing to Porto Novo at Haidar's request, de Suffren began to land the French troops on 21 February. Once the soldiers and their equipment were safely ashore, de Suffren set sail again in search of Hughes. The British were sighted off Ceylon on 9 April and battle was joined on the 12th. The French, as on 17 February, outnumbered Hughes; but, as on 17 February again, de Suffren was unable to make the most of his advantage. Dissent and jealousy amongst the French naval officers – indeed, outright failure to obey de Suffren's orders – vitiated the strength of the French squadron. Hughes escaped, and he was so relieved that he implied that the British had won a moral victory ('the Enemy's Squadron drew off in great disorder').[54] In truth, the engagement was as indecisive as most of the naval actions of the war.

On land, too, neither side was able to gain the upper hand, though in February 1782 Haidar's son Tipu wore down and then compelled the surrender near the Colaroon river in Tanjore of a British detachment under Colonel Braithwaite. To divert Haidar from the Carnatic, the Bombay authorities had sent a small force to Calicut on the Malabar coast. Haidar thereupon sent Tipu with part of his army to repel this incursion on his western territories. The British, having advanced inland, promptly retreated to the coast. To the east, meanwhile, the French were making only slow progress. They took Cuddalore on 8 April, but their commander, Duchemin, was too cautious to win the respect of Haidar Ali. Indeed, Haidar, outmanoeuvred by Coote at Arni on 2 June, was becoming so exasperated that he seriously contemplated coming to terms with the

British. To forestall this development de Suffren suggested a joint attack on Negapatam. The chief obstacle to the enterprise was Hughes's squadron, which anchored before Negapatam on 2 July. De Suffren offered battle four days later. The firing on both sides was very destructive – 'the combat was hot for some vessels', was how de Suffren reported it[55] – but after two hours a change of wind direction scattered the contending ships. In this sense the battle was inconclusive; Hughes, however, had achieved his objective in that Negapatam was saved.

The French now turned their attention to Trincomalee. Reinforcements under de Bussy had arrived at Isle de France, and de Suffren used part of this force for the attack. French soldiers, sepoys and Dutch auxiliaries were landed near Trincomalee on 26 August and the fort defending the port surrendered at the end of the month. Too late, Hughes arrived with the British fleet on 2 September. The next day de Suffren offered battle yet again. As in their previous engagements, French numerical superiority could not be brought to bear; for a time, indeed, Hughes was able to concentrate his firepower to out-gun the French centre. At length, however, nightfall obliged the weary and bloodied protagonists to break off the combat. The British returned to Madras to refit, and de Suffren sailed to northern Sumatra to winter his ships.

In December 1782 the Bombay government, no longer preoccupied with the Marathas, with whom terms had been concluded the previous May, sent a sizeable force under Brigadier-General Richard Matthews to attack Mangalore and the province of Bednur. When the British landed, they met little resistance because Tipu had returned to the Carnatic on hearing of the death of Haidar. Matthews took Bednur in January 1783 thanks to the treachery of one of Tipu's subordinates, and in March Mangalore surrendered to the British forces. In a re-run of the operations of the previous year, Tipu marched west from the Carnatic with the bulk of his army. In May he recaptured Bednur, and with it much of the army of the over-confident Matthews. The British garrison at Mangalore was laid under close siege. But if the British had been forced to yield most of the ground that they had gained in the west, Matthews had at least diverted Tipu from the Carnatic. Indeed, the French were even obliged to provide a contingent for Tipu's western expedition, leaving their force on the Coromandel coast, now commanded by d'Hoffelize, only 600 strong. Tipu agreed, however, to co-operate with de Bussy when he arrived with the main French force.

De Bussy disembarked with around 2,500 men at Porto Novo on 16 March 1783. He saw his primary task as restoring the morale of the French expeditionary force. He needed, as he put it, to bring the French troops 'out of the shameful langour in which Messieurs Duchemin and d'Hoffelize have allowed them to remain'.[56] But the French were largely left to their own devices, since even that portion of Tipu's army that remained in the Carnatic was employed elsewhere.

The British, recognizing the importance of dislodging de Bussy before Tipu returned from the west, started to move south from Madras on 21 April. Major-General James Stuart, who led the royal and company troops, made slow progress, and reached the Ponnaiyar River, near Cuddalore, on 2 June. Stuart successfully outflanked the French defences by a march to the west and by 7 June his army was south of Cuddalore. De Bussy re-deployed his troops to meet the British advance. On 13 June the British attacked part of the French lines, only to be met by a devastating fire from the well-entrenched defenders. The soldiers of a Hanoverian regiment and the newly-raised One-hundredth Foot retreated in disarray – 'never was there a greater panic than now prevail'd among the runaways'[57] – but when the French scrambled out of their works in pursuit, other British troops occupied a key redoubt. Honours were therefore even. The British had received a severe mauling, and the inexperience of many of their troops had been graphically revealed. On the other hand, the French defences had been breached, and de Bussy retired within the walls of Cuddalore.

De Suffren arrived on 15 June in time to save the French and, indeed, to tip the balance in their favour. Five days later Hughes and de Suffren fought their fifth and final battle. The British, reinforced by ships under Sir Richard Bickerton, now outnumbered the French. The action, as Hughes reported to the secretary of the admiralty board, was fierce: 'a heavy cannonade ensued, on both sides'. Several of Hughes's ships were 'much disabled in their Hulls Masts and Rigging'.[58] Eventually, the British withdrew to Madras. From the shore the French troops cheered, and it was generally thought, de Suffren wrote with understandable pride, that 'it was without precedent for 18 vessels to flee before 15'.[59]

Heartened by de Suffren's success, de Bussy launched an attack on his besiegers on 25 June. The French were repulsed – Stuart claimed that 'it was a most Complete route'[60] – but the British troops, deprived of naval cover by de Suffren and depleted in numbers, were now vulnerable. News of the signing of the peace treaties in Europe, which reached India at this crucial juncture, might well have saved Stuart from a damaging reverse. Lieutenant John Kennaway certainly had no confidence in the British position. The East India Company's agreement to suspend military operations, while having the appearance of 'an indecent Haste', was, in Kennaway's view, 'most judicious and wise' given Stuart's 'inadaquate Abilities' and the way in which Hughes had been 'obliged to forsake the Army'.[61] Tipu continued the fight into the new year, forcing the surrender of the garrison at Mangalore on 30 January; but without French help his chances of victory were greatly diminished. He finally agreed terms with Lord Macartney, the governor of Madras. By the peace of Mangalore (11 March 1784) both sides accepted a return to the prewar *status quo.*

*

The war beyond America was not, as some historians imply, just a series of minor side shows. For the British, the war in America diminished in importance once the Caribbean, Europe, India and Africa became arenas of conflict, as we can see when we compare the deployment of the British army before and after French intervention (see Table 1). The defence of Britain itself clearly became a priority once the Bourbons became open enemies, and troops were also despatched to the West Indies and India.

Table 1
Percentage of the British army in different theatres

	N. America	Britain	W. Indies	Mediterranean	India
Feb. 1778	65	26	2	7	0
Sep. 1780	29	55	8	6	1
% change	– 36	+29	+6	– 1	+1

Source: Piers Mackesey, *The War for America* (London, 1964), pp.524–5.

Similar considerations influenced the deployment of the Royal Navy. Early in 1778 there were 92 British warships of all types in North American waters. As soon as the British government learned of the signing of the Franco-American alliance the admiralty ordered 20 of Lord Howe's vessels to return home and a further 13 to go to the West Indies. Although at various times in the next five years the North American squadron was temporarily reinforced, both the West Indies and home defence took precedence (Table 2).

The reallocation of British military and naval resources caused by the broadening of the conflict naturally had important implications for the Anglo-American war. Some of these have been pointed out already, but they merit reiteration. Clinton bitterly resented having to release troops for service in the West Indies and complained that the reduction in the size of his army made the conduct of offensive operations very difficult. While he might have exaggerated his problems, his situation was certainly much less favourable than General Howe's in 1776 and 1777. In the spring and early summer of 1778 the British admiralty, fearing a French descent on the south coast of England, delayed sending ships to North America to support Admiral Howe against a powerful French squadron under d'Estaing. The result could easily have been a successful French blockade of New York and the destruction of Clinton's army. And, crucially, in the late summer and early autumn of 1781, when British naval strength in America might have been reinforced, the admiralty again felt obliged to retain a high proportion of the fleet in home waters to parry a possible Franco-Spanish invasion and to counter any Dutch challenge in the North Sea. A few extra British ships of the line in America might have saved Cornwallis's beleaguered army at Yorktown. As it was, without adequate naval cover, he was compelled to capitulate and the war in America effectively came to an end.

Table 2
Percentage of British ships of the line in different theatres*

	Europe	North America	West Indies	India
July 1778	47	41	8	3
July 1779	49	9	33	9
July 1780	34	13	41	8
April 1781	38	11	36	5
April 1782	21	10	48	18

* All figures include ships *en route* to, returning from, and fitting for the theatre in question
Source: Jonathan R. Dull, *The French Navy and American Independence* (Princeton, N.J., 1975), pp.359–76.

Chapter 6 Notes

1. *Extracts from the Diary of Christopher Marshall*, ed. William Duane (Albany, N.Y., 1877), p.277.
2. R.B. Sheridan, 'The Wealth of Jamaica in the 18th Century', *Economic History Review*, 2nd series, xviii (1965), 292–311.
3. Philip R.P. Coelho, 'The Profitability of Imperialism: the British Experience in the West Indies, 1768–1772', *Explorations in Economic History*, x (1973), 253–80. See also R.P. Thomas, 'The Sugar Colonies of the Old Empire: Profit or Loss for Great Britain?', *Economic History Review*, 2nd series, xxi (1968), 30–45; and Sheridan's 'The Wealth of Jamaica ... a Rejoinder', ibid., 46–61.

4. *The Lost War: Letters from British Officers during the American Revolution*, ed. Marion Balderstone and David Syrett (New York, 1975), p.211.
5. Alexander, Lord Lindsay, *Lives of the Lindsays* (4 vols., Wigan, 1840), iii. 220.
6. Piers Mackesy, *The War for America, 1775–1783* (London, 1964), p.232.
7. Grant to Gov. William Burt of St. Kitts, 22 April 1779, CO 318/5, fo. 82.
8. Lt. William Robertson to Duncan Robertson, 17 Jan. –1 Feb. 1779, Dalguise Muniments, GD 38/2/19, SRO.
9. Archives de la Marine, B³ 661, fo. 279, Archives Nationales.
10. Byron to Philip Stephens, 8 July 1779, ADM 1/312, fo. 89.
11. CO 318/5, fo. 100.
12. *The Diary of the Revd. William Jones 1777–1821*, ed. O.F. Christie (London, 1929), p.58.
13. CO 137/75, fo. 200.
14. Rodney to Stephens, 26 April 1780, ADM 1/311, fo. 157.
15. Broughton and Cally Muniments, GD 10/1421/7/338, SRO.
16. CO 318/6, fo. 79.
17. *Sandwich Papers*, iv. 148.
18. Hood to Rodney, 4 May 1781, Rodney Papers, PRO 30/20/12.
19. Lt.-Col. James Cockburn to Brig.-Gen. Gabriel Christie, 27 Nov. 1781, Egerton MS 2135, fo. 197, British Library.
20. Journal of Midshipman James Francis Grant, Add. MS 45,124, fo. 47.
21. Archives de la Marine, B³ 716, fos. 123–4.
22. *Letters and Papers of Charles, Lord Barham*, ed. Sir John Knox Laughton (3 vols., London, 1907-11), i. 178.
23. *The Naval Campaigns of Count de Grasse*, ed. and trans. Amandus Johnson (Philadelphia, 1942), p.103.
24. *Sandwich Papers*, iv. 257.
25. Hood to Capt. Francis Reynolds, 2 Jan. 1783, Ducie and Morton Muniments, D 340a C28/7, Gloucestershire RO.
26. For more positive assessments of Sandwich's thinking, which stress the value of concentration in home waters, see David Syrett, 'Home Waters or America? The Dilemma of British Naval Strategy in 1778', *The Mariner's Mirror*, lxxvii (1991), 365–77; and N.A.M. Rodger, *The Insatiable Earl: A Life of John Montagu, 4th Earl of Sandwich* (London, 1993), pp.271–9.
27. Keppel to Stephens, 30 July 1778, ADM 1/94, fo. 461
28. *The Lost War*, p.165.
29. Entry of 19 Aug. 1779, Diary of Richard Neville, Neville and Aldworth Papers, D/EN F55/1, Berkshire RO.
30. *Sandwich Papers*, iii. 74.
31. Drake of Buckland Abbey Papers, 346M/F23, Devon RO.
32. Add. MS 46,473, fo. 86.
33. Eliott to Viscount Weymouth, 25 Aug. 1779, CO 91/25.
34. ADM 1/311, fo. 50.
35. Archives de la Marine, B³ 698, fo. 339 (intercepted letter).
36. Add. MS 38,605, fo. 3.
37. Eliott to [Amherst,] 27 May 1781, WO 34/133, fo. 3.
38. Ducie and Morton Muniments, D 340a C30, Gloucestershire RO.

39. *Archives ou correspondence inédite de la maison d'Orange-Nassau*, ed. F.J.L. Krämer, 5th series (3 vols., Leiden, 1910–15), ii. 607.
40. Cuninghame of Thorntoun Muniments, GD 21/629/7, SRO.
41. CO 174/14, fo. 7 (12 Nov. 1781).
42. Archives de la Marine, B^3 731, fo. 14.
43. *Spain under the Bourbons*, ed. and trans. W.N. Hargraves-Mawdsley (London, 1973), p.162.
44. Add. MS 38,605, fo. 142.
45. *A History of the Late Siege of Gibraltar* (London, 1785), p.291.
46. *Private Sea Journals 1778–1782 kept by Admiral Sir Thomas Pasley, Bart.*, ed. Rodney M.S. Pasley (London, 1931), p.142.
47. Sir Hector Munro to Lord Weymouth, 27 Oct. 1778, Home Misc./142, p. 53, IOL.
48. Mackenzie to ..., 3 Nov. 1780, Cromartie Papers, Centre of South Asian Studies.
49. Thomas Munro to Alexander Munro, 11 Oct. 1780, Munro Collection, MSS Eur. F. 151/140, IOL.
50. John Cleveland Jr. to John Cleveland, 15 Oct. 1780, Benenden Letters, Add. MS 41,665, fos. 14–15.
51. Diary of an unidentified officer, MSS Eur. C. 156a, IOL.
52. Coote to Lord Macartney, 4 July 1781, Macartney Papers, Add. MS 22,439, fo. 31.
53. Munro to Lord Hillsborough, 24 Nov. 1781, Home Misc./160, p.524.
54. Hughes to Philip Stephens, 10 May 1782, ADM 1/164, fo. 298.
55. *La campagne des Indes: lettres inédites du Bailli de Suffren*, ed. Régine Pernoud (Mantes, 1941), p.60.
56. Quoted in S.P. Sen, *The French in India 1763–1816* (Delhi, 1971), p.318.
57. Diary of an unidentified officer, MSS Eur. C. 156a.
58. ADM 1/166, fo. 77.
59. *La campagne des Indes*, p.71.
60. Stuart to Lord Macartney, 25 June 1783, Stuart Stevenson Papers, MS 8345, fo. 194, National Library of Scotland.
61. Kennaway to Sir Robert Palk, 13 July 1783, Private Letter-book 1782–5, Kennaway Papers, 961M/F1, Devon RO.

7

The Impact of the War on America

'The whole country gives evidence of the horrors of war', wrote a French officer while passing through the 'neutral ground' near British-held New York City in July 1781. The houses were 'deserted, half destroyed, or burned, with untended orchards and gardens filled with fruits and vegetables, and driveways overgrown with grass 2 feet high'.[1] More than three years earlier, a Scottish officer in the British army had been similarly appalled by the condition of the farmland near the British base in Philadelphia: 'this once happy Country exhibits a scene of desolation to[o] horrible for Language to express'.[2] Robert Honeyman, a Virginian doctor, recorded less emotionally the awful appearance of a local estate after the British troops had encamped there in June 1781: 'The fences pulled down & much of them burnt; Many cattle, hogs, sheep & poultry of all sorts killed ... & the offal ... &c with dead horses & pieces of flesh all in a putrefying state scattered over the plantation'.[3]

These various accounts of the physical impression left by the conflict remind us, of course, of only one facet of the impact of the war. They are perhaps unrepresentative in another sense. They relate only to localities that experienced significant military operations. There were whole areas of the new United States that escaped lightly or were untouched by the ravages of war. But even in these areas nearly every American, however much he or she tried to continue with their daily routine and ignore the conflict, was affected in one way or another. In Marshfield, Massachusetts, which after the initial fighting was remote from military operations, Sarah Thomas lamented in March 1779 'the want of bread', and prayed that God would 'hasten the time when we shall again have peace in our once happy land – and the P[e]ople which are now imployd in destroying one another Spend there time in Cultivating the ground that there poor unhappy Famalys may have bread to eat'.[4] To explore the war's impact in a comprehensive manner would require a great deal more space than can be devoted to it here. What follows must therefore be merely suggestive.

*

The war had some immediately beneficial effects on the American economy. We should recognize the ability of the new nation, freed from the constraints of the Navigation Acts, to export staple crops directly to European markets: Chesapeake tobacco could now be sold to France without the services of the Glaswegian middlemen whose charges increased the cost to French consumers and so reduced the profits for American producers. Indeed, tobacco exports played a crucial part in sustaining the American war effort: a British army lieutenant employed in burning tobacco stocks in Virginia in 1781 justified this incendiary activity by saying that 'it was the staple commodity with which they [the revolutionaries] procured warlike stores'.[5] Overseas commerce was of course disrupted – New England fish exports were severely diminished by restricted access to the Newfoundland cod banks, and British markets were closed to American goods of all kinds – but a significant volume of produce found its way to new trading partners in Europe and the Caribbean. The foreign West Indian islands were important, too, as staging posts in the American import trade. Foreign goods – including British manufactures sent first to Europe – were despatched to the West Indies and there exchanged for American exports. St. Eustatius, as we have seen, played a central role in this wartime commerce; and Spanish Cuba was another Caribbean entrepôt, particularly favoured by merchants from Philadelphia and Baltimore. Privateering profits might also be considered as a wartime benefit: American privateers captured British vessels to the value of around £18 million sterling. Loans and subsidies provided by the European allies were a definite plus: they proved useful in helping to offset the imbalance in the commodity trade, which, with exports worse hit than imports, was more severe than before the conflict began. Between 1777 and 1783 these loans amounted to the equivalent of more than $10 million, and in 1781 France alone provided 20 million *livres* or $3.7 million.

Internally, there were certain infrastructure improvements. These ranged from the construction of new roads to facilitate the movement of military supplies to the foundation in 1781 of the Bank of North America, based on the credit of the Philadelphia merchant and war profiteer Robert Morris, who was appointed by Congress as superintendant of finance. More important was the stimulation given to the relatively-underdeveloped domestic sector by the fall in imports. Embargoes on British products had already encouraged American import substitution before the fighting (the

Continental Association had been operating since the autumn of 1774) and the sharp reduction in imports of manufactured goods during the war continued to have a stimulating effect. The main beneficiaries were the fledgling textile, paper, pottery, and shoe industries, together with more developed iron and steel manufacturing. Cereal farmers should be included, too. Aided by the reduced imports of high-quality West Indian rum, and the molasses used for making the lower-grade New England version, they found that their crops were in great demand for the production of beer and whisky as replacements.

The paper money issued by Congress and some of the states further quickened the pace of economic activity at home; and, until late 1776 at any rate, the money supply did not increase beyond what was needed to service an expanding economy. Military demands similarly boosted domestic production. The need for footwear, uniforms, tents, munitions and other military stores helped American shoemakers, gunpowder manufacturers, gun-founders and textile producers. Small wonder that in September 1781 Philadelphia merchants told French officers passing through the city that they had no desire for peace, because peace would damage their business. A good many American farmers probably shared this sentiment: the Continental and French armies required vast amounts of food; and even the British, who brought salted provisions across the Atlantic, were always anxious to obtain fresh local supplies to keep up the morale of their troops. The French and the British, moreover, had the necessary specie to attract farmers willing to sell. French military expenditure in 1782 was said to have amounted to the equivalent of around £1 million sterling; and the British contribution was probably greater. We know that between 1 January 1776 and 8 October 1783 £13,166,664 was disbursed by the paymaster-general in London to the company of Harley and Drummond for the supply of money for the use of the army in America. In other words, as much as £1,645,830 a year could have been injected into the American economy.

Nor should we assume, as some commentators seem to have done, that the impact of this money was felt only in the British enclaves.[6] True, there was a noticeable rise in prices in the occupied towns – an eightfold increase in New York City between 1776 and 1783 – which could be explained by the great augmentation of the quantity of money in circulation. And trade in the British-held areas was certainly brisk – another indication, perhaps, that they, rather than the colonies generally, were the gainers by the spending power of the royal forces. New York had experienced nothing like it since the golden days of the Seven Years' War. William D. Faulkner, a city brewer, regularly 'Sold to Major McLaraths Mess' gallon upon gallon of beer; but Faulkener supplied many other regiments besides McLeroth's – British, Hessian and loyalist provincial – and he also had a contract with the commissary general ('Six Shillings for every 100 Gall^{s}.')[7] The account-books of New York merchants and shopkeepers similarly testify to buoyant sales; in Charleston the mushrooming of retail outlets provoked comment; while in Philadelphia Thomas Clifford Jnr. observed: 'the Army of itself takes off large Quantys of Goods & by giving Employ to many of the Inhabitants enables them to become Purchasers also, by which means there has been Considerable Sales'.[8]

But the beneficial effects of British military spending were not confined to the areas held by the British. The defensive lines encircling the occupied towns were designed to repel American attack, not to keep out commodities from the revolutionary hinterland. Goods from adjacent areas flowed into the British enclaves, where they were exchanged for British manufactures or for money. From newly-taken Charleston James Simpson reported that, lured by the promise of specie payments, colonists were travelling great distances to sell their produce. 'Wagons with tobacco', Simpson explained, 'have come upwards of two hundred and thirty miles'.[9] Farmers in Pennsylvania and New Jersey likewise brought goods into British-held Philadelphia – despite Washington's attempts to stop them – and the British base at New York City drew food supplies from New Jersey to the west and New England to the east, as well as from those parts of New York State outside British control. And the British, of course, did not remain closeted in their bases. In their peripatetic progress through the colonies they simply seized many of their requirements from colonists whom they regarded as enemies; but they paid for some goods or services, and sometimes paid handsomely. Few Americans can have been as lucky as Gerard W. Beekman. When Clinton's army evacuated Philadelphia and slowly traversed New Jersey on the way to the safety of New York City, Beekman was obliged to part with two horses. He received £47 5s hard currency, which he considered 'No Grate Loss as they had Both be[e]n Foundered Verey Bad'.[10] But whether Americans whose property was requisitioned received more or less than their deserts, many of them received some British money, adding to the quantity that found its way into the territory beyond the British lines.

So much for the positive side. Now we should look at the negative. If escape from the Navigation Acts brought some advantages, the collapse of British mercantile regulation was a mixed blessing. The loss of the subsidy provided by Parliament for indigo cultivation in South Carolina had an immediate impact: indigo growing all but ceased. Americans may have captured many British vessels, but the British almost wiped out the overall benefit to the American economy by seizing nearly as many. John

Kirby, a sailor on HMS *Unicorn,* told his father in October 1778 that, since their arrival in American waters, the crew had captured 'abought 50. Saile of Vessels', most of which were merchantmen.[11] The activities of the Royal Navy and British and loyalist privateers naturally put upward pressure on insurance rates – both in America and Europe – and as these rose the cost of overseas trade increased. The fall in imports, furthermore, fuelled price inflation. The quantity of foreign imports on the market was nearly always well below demand. Despite the Continental Association – perhaps because of it – Americans were anxious to get their hands on British manufactures in particular. Internal transport disruption, particularly the requisitioning of large numbers of wagons and carts by the competing armies, added to the problem by increasing the cost of home-produced as well as imported goods.

The production and price of American products were also affected by the flight of large numbers of slaves who sought the protection of the British: Robert Taylor of Virginia was just one of countless slave owners who complained that when the British army arrived in his neighbourhood 'Several of my Negroes ran away from me'.[12] Indentured servants were nearly as keen to escape. Lord George Germain's recommendation in 1778 that they be given every encouragement to join the British in order to undermine the American economy was unnecessary: as Howe's army advanced across south-east Pennsylvania towards Philadelphia in September 1777 indentured servants deserted their masters in droves (Thomas Waters of Chester County, for example, recorded the loss of 'a servant that had about thirteen month to Serve').[13] The result, inevitably, was reduced output and higher prices. When free workers left the labour force in large numbers to serve in the Continental army, or loyalist provincials, or militia, an added problem was upward pressure on wages. William Whipple of New Hampshire argued that privateering had the same effect: the five privateers being fitted out in Portsmouth would, he reckoned in July 1778, 'take four hundred men', mainly 'countrymen',[14] and so would push up inland wages and therefore provision prices.

Inflation can mainly be attributed, however, to the printing of vast quantities of paper money in which few Americans had much confidence. We have already seen that by 1780 Congress and the states between them had issued some $440 million in bills. This rapid expansion of the money supply was bound to have a dramatic effect on monetary values and therefore on the prices charged by producers and retailers. In Pennsylvania prices rose by 45% in a single month of 1779, and by the end of that year they were seven times higher than at the beginning. For Marylanders the cost of such essentials as beef, butter, corn, wheat and salt increased by between 1,900 and 5,043% in the period 1777–80. Consumers, particularly urban consumers, complained bitterly, especially as wage rises fell well behind commodity prices. They pressed for price controls, which were adopted by the revolutionary and British authorities but largely failed to stem the tide. In the areas under revolutionary jurisdiction there was to be no real improvement until deflationary policies were adopted by Congress and the states in 1781–3. Then the producers, who had often gained from the paper money system, saw their bloated profits evaporate and became as vociferous in their protests as consumers had been before. Hyperinflation and its antidotes proved to be as divisive and damaging to American morale as almost any other wartime development.

If rampant inflation is one of the most well known economic consequences of the war in America, we should not forget that the physical destruction or removal of property was no less important. Mention has already been made of the loss of merchant shipping to the Royal Navy and British and loyalist privateers. The British forces also raided coastal towns, from Falmouth (now Portland) in modern-day Maine in October 1775 to New London and Groton, Connecticut, in September 1781. Not that inland areas were necessarily safe: from the Cheraw district of South Carolina Major James Wemyss reported in 1780 that he had 'burnt & laid waste about 50 Houses & Plantations';[15] and frontier areas experienced perhaps even worse devastation at the hands of British soldiers and their Indian auxiliaries. After a particularly notorious incursion in July 1778, Major John Butler proudly listed his achievements: 'we have taken and destroyed eight pallisadered Forts, and burned about 1,000 Dwelling Houses, all their Mills ... we have also killed and drove off about 1,000 head of horned Cattle, and sheep and swine in great numbers'.[16]

Many British officers, we have seen, were keen to pursue a war of destruction to terrorize Americans into submission. But domestic and commercial property was regularly destroyed or damaged for other reasons. When the army required materials for the construction of fortifications, it was only to be expected that wooden houses and outbuildings would be used. Silas Cooke of Rhode Island lost the floor boards from two of his still-houses when the British required timber to construct gun platforms during the siege of Newport; while Francis Blackburn, a clothier and fuller of Oysterbay, Long Island, was 'intirely Deprieved ... from following his Business' after his mill and equipment were destroyed by British troops seeking timber to build barracks.[17] Military necessity similarly led to the demolition of buildings to provide defenders of fortified posts with a clear field of fire and to deny shelter to the enemy.

Other Americans, whose buildings remained standing, lost many or all of their contents. The letters and claims of Americans of every political persuasion bear witness to the range of goods stolen by British troops.

Robert Pleasants, a Virginia Quaker, wrote of how his son had been robbed of 'almost all the Cloth[e]s he had besides what was on his Back'.[18] William Hooper of North Carolina related how his home had been stripped of 'every article of household and kitchen furniture, knives, forks, plates, and spoons – an almost general sweep'.[19] Other accounts mention women's clothes, silver and gold items, books, and, most frequently, drink. Goods that were not taken by marauding soldiers might be officially requisitioned. There were, as we just saw in the case of Gerard W. Beekman, some Americans who gained by this practice. Many, however, were losers. Requisitioned goods were rarely paid for on the spot. Those lucky enough to be promised payment did not always receive anything; and those who were paid often complained that they were cheated. The people of Westchester County, New York, were said to be furious at receiving no more than 'about half Price' for the crops seized from them by the king's troops.[20]

We can only guess at the sums involved in all these American losses. We know that during the siege of Boston the inhabitants were believed to have lost property to the value of £17,620. But this omits loyalist losses, which by all acounts were considerable. Claims submitted at the end of the war by the citizens of York County, Virginia, suggest that the British army caused damage to the tune of more than £26,000. In 1781, when New London and Groton were put to the torch, the total losses were computed at £61,400 4s 6d. The Pennsylvania assembly reckoned that the British caused nearly £200,000 worth of damage in and around Philadelphia during their occupation of the city. Abraham Hasbrouk, one of the victims of General Vaughan's attack on Esopus, thought that he alone had lost property worth 'Between £900 and 1000 pounds';[21] while John Lloyd, a South Carolina planter, estimated that his own estate had suffered 'to the sum of near £8,000 Sterling' and that 'The property that the Inhabitants of this State have lost by means of the British ... amounts to upwards of £3,000,000 Ster^{g}.'[22]

And all this, we should remember, takes no account of the damage and destruction caused by the allied armies. The French were mainly kept away from temptation and were able to pay for most of their requirements; but the Continental army and revolutionary militia could be as destructively undisciplined as the British: Ebenezer Hazard, riding through New Jersey in August 1777, noted that Princeton 'looks like the Picture of Desolation', but added that the damage done to the college was 'more from the Licentiousness of our own troops than from the ravages of the Enemy'.[23] And, as with the British, even well-behaved soldiers could cause problems. American troops requisitioned goods, often offering owners little or no payment. Lieutenant Enos Reeves of the Pennsylvania line observed in November 1780 how an inhabitant was 'a little mortified at the army coming to winter so near'. In the last winter, Reeves explained sympathetically, the American forces had 'destroyed six hundred acres of woodland for him and his step-father, and I suppose this Winter [we] will clear the plantation of every tree'.[24]

*

In the pre-war years American society was far from united and harmonious. There was evidence of class antagonism between artisans and merchants in some of the towns and between tenants and landlords in those areas where land ownership was concentrated in a few hands; there were clashes between the long-established tidewater and the newer frontier communities; there were religious battles between and even within denominations; and there was ethnic friction – not just white against black, but between different types of white settlers. In some cases the war cut across these divisions, uniting diverse groups either for or against American independence. But the conflict could just as easily deepen as heal existing divisions.

We have already seen how the British army exacerbated the well established tension between the Scotch-Irish frontiersmen and the German farmers in the South Carolina backcountry. With the appearance of British troops in the early summer of 1780, the Germans started to assert their loyalty and to look to the British for protection. This seems to have been sufficient to arouse the resentment of the neighbouring Scotch-Irish, who now saw the Germans in an even more unfavourable light. The Scotch-Irish began to offer violent resistance to Cornwallis at least partly because they saw him arming and organizing the Germans. But the intensification of existing hostility was nowhere more obvious than in the Hackensack Valley of New Jersey. The local community was predominantly Dutch and was riven by a long-standing controversy within the Dutch Reformed Church. The conservative *conferentie* faction favoured continued adherence to the ways of the early-American Dutch Church, while the *coetus* group was a product of the evangelical fervour associated with the mid-eighteenth-century Great Awakening. The mutual antagonism remained within bounds until the British army arrived in late November 1776. Some of the *conferentie* were unable to resist the temptation to settle old scores by encouraging British troops to plunder the homes of their opponents. Now it is true that many of the *coetus* party were already sympathetic to the patriot cause, just as the *conferentie* faction was temperamentally disposed to loyalism. But there can be little doubt that the war aggravated the conflict between the two groups, giving it a new and dangerous dimension.

In some cases white hostility to blacks was also intensified by the war. In the south, where the slave population was particularly large – blacks outnumbered whites in several counties of North Carolina and Virginia and formed an absolute majority in South Carolina – the spectre of a large-scale slave uprising haunted many whites. Once the fighting commenced, southern whites became convinced that the British would sponsor such an uprising to undermine American resistance. As early as June 1775 a Charleston merchant wrote of the threat of a British-inspired 'insurrection from our slaves'.[25] The following November Lord Dunmore, the royal governor of Virginia, gave substance to these fears by offering freedom to all patriot-owned slaves who flocked to his standard. Clinton renewed the promise in 1779. No such encouragement was needed. Blacks were already leaving for the nearest British base, irrespective of the political allegiance of their owners. Anxiety that the British might arm these escapees was less intense in the north, where slaves formed a much smaller proportion of the population; but here owners who prided themselves on their paternalistic care could feel a deep sense of betrayal. When William Beekman wrote of the flight of his slaves Hanover and John in December 1776, he described them as 'Ungratefull' and vowed that he would never 'have either of them under my Roof again'.[26]

From the black perspective, of course, the war seemed like a great opportunity. In reality, slaves all too often found that they had swapped one form of servitude for another. William Carson of South Carolina discovered one of his runaways in the British camp near Charleston, acting as a footman to the wife of a common soldier. Some slaves who joined British raiding parties were promptly sold by their 'liberators'. Even so, the conflict brought long-term advantages for others. A number of whites recognized that the revolutionary rhetoric of liberty sat uneasily alongside the continuing institution of chattel slavery. In the north blacks themselves even used the courts to force whites to live up to their fine words. Black military service on the side of the revolutionaries was also important in helping to modify white American attitudes. Southern whites might have been reluctant to arm any blacks, whether free or unfree, but in the north, as we have seen, blacks were conspicuous in Continental army units. This, as much as revolutionary ideology, probably explains the softening of the official line on slavery in the northern states. Their legislatures, admittedly, were cautious and indirect in their approach. Massachusetts merely dropped any reference to slavery in its constitution of 1780. Connecticut prohibited the sale of slaves outside the state in 1779, but said nothing explicit about abolition. Only Vermont, a new state, banned slavery outright, proclaiming its stance in the constitution of 1777. But despite this general circumspection, a new climate of opinion was clearly beginning to form. Further south, slave-owners whose blacks had resisted British blandishments during the war increasingly chose to express their gratitude by individual acts of manumission. In the Chesapeake the number of free blacks rose from a few thousand in 1776 to 60,000 in 1810.

For white American women, too, the war brought both problems and opportunities. On the negative side we have to include all the disruptions caused by hostilities. Women were often separated from husbands, brothers and fathers serving in the armed forces. As Elizabeth Ambler wrote in 1781, the war was terrible 'however distant', but seemed even more so when 'those we most love are personally engaged in it'.[27] For a good number, moreover, the separation was to be permanent: according to one estimate, a higher proportion of American soldiers was killed than in any other conflict except the civil war of 1861–5. And when the war came near to home, women had added cause for dismay. While Sally Wister, a Pennsylvanian teenager, may have found nothing to complain about when American troops were quartered on her family's temporary home, other women regarded the new arrivals with less equanimity. Sarah Logan Fisher, a Pennsylvanian loyalist, berated the revolutionary soldiers quartered at Germantown for their ill-conduct: they were, she wrote, no better than 'a lawless banditti'.[28] But however badly the American forces behaved, they were generally less feared than the British. 'Oh how I dreaded the approaching enemy!', Eliza Wilkinson recalled.[29] Cornelia Bell of New Jersey was equally alarmed: 'Thank Heaven I have seen none of them yet and hope I never shall, though we have been in daily expectation of them for some time past'.[30] What lay at the back of these women's minds, of course, was the image of a drunken and libidinous soldiery: women through the ages have had special reason to fear the arrival of a conquering army. Revolutionary accounts naturally played up the number of sexual offences committed by British troops – 'The Devils ... have waged War against Decency', wrote an outraged American officer;[31] 'There is Scarcely a Virgin to be found in the part of the Country that they have pass'd thro', a congressman claimed[32] – but even British records imply that the notoriety of their soldiers was deserved.

More positively, we can see that, in the absence of so many men, new opportunities arose for women. By the middle of 1776 some 4,000 were working in the environs of Philadelphia in the domestic textile industry – an industry that was given a very large boost by the war. In the city itself, women made door-to-door collections to raise money to help the Continental army. There was, to be sure, a traditional side to this enterprise; women, after all, were still cast in a supportive role. But the collections, and the committees established to organize them, were directed and led by women themselves. They seem to symbolize a female desire to participate in the world of politics. One should not exaggerate the impact of such limited assertiveness; but even when women were not consciously seeking a new status the war could thrust it upon them. Of necessity wives had to take more of an interest in matters beyond their normal sphere while their menfolk were serving in the armed forces, or were otherwise taken from their homes. James Pemberton, a Quaker exiled from Philadelphia by the revolutionary authorities, was told by his son that 'Mother thought proper to cut all the Wood down' on their estate to hide and protect it from marauding British and Hessian soldiers.[33] Mrs Pemberton had been forced to take a decision that her husband would surely have made in more normal circumstances; and, emboldened by the experience, she wrote to a British officer to complain of his undisciplined behaviour. Years later, the more humble Azubah Norton of Connecticut remembered that the frequent absences of her soldier-husband Benjamin meant that he 'was unable to do anything on our farm. What was done, was done myself'. With this broadening of female responsibilities came a greater male recognition of female worth – in some instances, at any rate. As Colonel Timothy Pickering told his wife Rebecca in August 1783: 'This war which has so often & long separated us, has taught me how to value you'.[34]

The war had an effect on land ownership, though perhaps not as dramatic an effect as was once believed. It used to be assumed that the departure of many prominent loyalists, and the confiscation of their property, led to a much wider distribution of land ownership in the new republic than under the old colonial regime. There were certainly some changes. Nearly 200 tenants obtained freehold status, for instance, on the 50,000 acre Phillipse estate in Westchester County, New York; and about the same number in neighbouring Dutchess County, where the enormous manors belonging to the loyalists Beverley Robinson and Roger Morris were broken up. In Maryland, another province with a large pre-war tenant population, there was also a noticeable transfer of property to former lease-holders. But historians are now much more cautious about advancing theories of a social revolution based on a more democratic pattern of land ownership. Part of this caution derives from a realization that the loyalists were not – as older accounts tended to imply – exclusively or even primarily members of the social elite. Loyalists were as likely to be artisans or small farmers. Some were tenants who fought in loyalist provincial corps raised by their landlords and who followed those same landlords into exile. More important, we now know that much of the prime developed land that was sold during or after the war was purchased by existing land owners. It was not unusual for already-substantial estates simply to become bigger still. In New Jersey, according to a recent study, the general pattern of land-holding remained more or less unchanged, with tenancy as pervasive as it had been before the revolution.[35]

None the less, the war did make some difference. Emphasis is now placed on the benefits that flowed from the removal of British restrictions on westward expansion, combined with the decline in the Indian threat on the frontier once the British had been defeated. The opening up of new lands is seen as promoting social mobility by allowing hitherto impoverished and landless Americans to acquire their own farms. Former soldiers were even given land on favourable terms as a reward for their services and to encourage western settlement. As a result, pressure was reduced on the increasingly densely-populated eastern areas – particularly long-established New England – which helped to halt the trend towards a more European model of society which was becoming obvious in the last decades of the colonial period, a model characterized by greater concentration of wealth and more social differentiation.[36]

The war, indeed, might well have put this process of creeping Europeanization in reverse; or rather, perhaps, set American society moving in an entirely different direction. During the revolutionary period America was transformed, Gordon S. Wood tells us, from a society that was 'monarchical' and 'hierarchy-ridden' into one that was democratic and 'commercially minded'. The war years played their part in this transformation. New opportunities for land ownership no doubt furthered a more egalitarian spirit; but there were other, perhaps more important, contributors. One was the wartime boost to the domestic economy, which, to quote Wood again, 'brought new producers and consumers into the market economy, ... aroused latent acquisitive instincts everywhere, and ... stimulated inland trade as never before'.[37] Revolutionary leaders – and in all probability many of their followers – who often portrayed themselves as engaged in a struggle to defend a simple, rural, and virtuous society that was under threat from the British, might decry this commercialization. Nathanael Greene's correspondence, for example, is full of lamentations at the decline in public spirit and the prevalence of selfishness and profiteering. But, unwelcome as these developments might have been to those who clung to traditional ideas of public virtue, the values of the market-place were, in their way, perhaps even more egalitarian than the communitarian values of the early settlers. Popular materialism certainly undermined established ideas of hierarchy. Money – however obtained – seemed to be eclipsing birth. The comments of one of

Rochambeau's officers, a Swedish nobleman, are instructive in this regard: 'money is their god', he wrote disdainfully of the Americans and their willingness to trade with the enemy; 'virtue, honour, seem nothing to them compared to the precious metal'.[38]

Another relevant development was surely the humbling of a portion of the old colonial elite. Respect for rank must have been undermined by the fall from grace of powerful loyalists, whether land owners, merchants, or royal officials. We should also note the exercise of political power – on local committees, even in state assemblies – by men of humble origin, which was both a cause and a symptom of a greater sense of equality. Political power had not, of course, been monopolized by the elite in the colonial period. In this respect, colonial America was already more egalitarian than eighteenth-century Britain. Abraham Yates Jr., an artisan from Albany, New York, was able to hold the office of county sheriff long before 1775. But that was as far as Yates could reasonably expect to go. It took the Revolution and the war to allow him to become a United States congressman. As an amazed French officer noted in 1781: 'a locksmith, a cobbler, or a merchant may become a member of Congress'.[39] This political advance of the lower ranks of society was perhaps most conspicuous in Pennsylvania, where the colonial old guard was toppled by the radical artisans of Philadelphia. Members of the traditional elite were acutely aware that a dramatic change had occurred. The son of the former chief justice remarked bitterly in January 1777, 'the most insignificant now lord it with impunity & without discretion over the most respectable characters'.[40] Dr. Benjamin Rush, a revolutionary but an opponent of the new state constitution of 1776 that enfranchised nearly all adult males, was still more scathing: 'They call it a democracy – a mobocracy in my opinion would be more proper. All our laws breathe the spirit of town meetings and porter shops.'[41]

Popular military participation was as important as popular political involvement; indeed, in many places they went hand in hand. The radical Philadelphia artisans were soldier-citizens prepared to defend the new order, which is essentially why the old elite had to give way and accept such a democratic state constitution. Elsewhere, if less spectacularly, military service also furthered a more egalitarian, or more meritocratic, society. Even in the Continental army, which had plenty of officers taken straight from the top drawer of colonial society, merit jostled successfully with birth. As the war dragged on, more and more Continental officers were promoted from the ranks. But the military contribution to social levelling was still more evident in the revolutionary militia. Prosperous Americans could find themselves serving as common soldiers alongside those whom they were accustomed to view as their social inferiors. In

Pennsylvania, Captain James Young's company of the Cumberland County militia included the likes of Henry Coyle, a doctor and landowner, Patrick Campbell, a merchant, and George Cunningham, a farmer who owned 150 acres. These local worthies saw duty with Nicholas Swope, a sadler, John Beatty, a blacksmith, and John Taylor, a tailor, and among their company sergeants was William Shanon, a stay-maker.[42] And, as we have seen, the officers in militia units – especially in the north – might be elected by their fellows rather than appointed by state officials, and might therefore be men of no great status; of less social standing, indeed, than some of those whom they commanded. The experience of this kind of military service – particularly in the revolutionary militia – was almost bound to have eroded social barriers and furthered a sense of equality.

A sense of American-ness, a greater awareness of the nation, seems to have been another product of the conflict. One of the causes of this was perhaps the greater geographical mobility of Americans during the war, many seeing parts of the new United States that they had never visited before. Some of the movement, it is true, was distinctly local. Elizabeth Hadwen and her parents left Newport, Rhode Island, early in 1775 'as there was great threats from the Men of War of firing the Town';[43] but they only went to Portsmouth, a mere nine miles away. Others, however, did their best to put a great distance between themselves and the fighting; time and again British officers commented on the way in which the inhabitants abandoned their homes on the approach of the army. And if non-combatants were uprooted, American soldiers often travelled much farther. Militiamen might serve in neighbouring states – we have seen how units from North Carolina and Virginia fought in South Carolina in 1780 – and Continental regiments were frequently engaged in campaigns far from home. Troops from Pennsylvania were involved in the South Carolina operations in 1781; men from Massachusetts fought in Virginia at the same time; the Delaware Regiment was engaged in the battles in New York and New Jersey in 1776–7. Nor was it just a matter of the temporary widening of horizons: inter-state marriages increased significantly, in part, at least, because young American males in uniform were on the move.

But as an agency for the promotion of national consciousness, the Continental army had an importance that transcended the experiences of individual soldiers and their partners. It helped to increase national feeling among Americans in general. Like Congress itself, the army was the embodiment of the new United States: it contained soldiers from the different states and was not answerable to any single state. And, more clearly than Congress, it was the vital prerequisite for the survival of the

Revolution. So long as the Continental army remained in being, so did the republican cause. Americans of neutral disposition might refer even-handedly to 'the Inglish army' and 'the americans',[44] but to those who supported the Revolution, the Continental army came in time to symbolize the new nation. It became 'our army', or simply 'the army', and its fortunes were followed closely by people the length and breadth of the nation. Americans might have worried about the apparent contradiction of relying on regular troops to secure their independence (after all, they had grown accustomed to regarding standing armies as a threat to liberty); they might have preferred, for this reason, to fight the British with a militia army of citizen-soldiers; but they soon realized that the Continental army was indispensible: without it they could not hope for deliverance.

It might reasonably be objected that provincialism remained strong. Most people probably still felt a greater loyalty to, say, Maryland, or New Hampshire, than to the United States. This persistent provincialism found expression in the constitutional arrangements designed to link the states: after having resisted parliamentary authority for so long, Americans were understandably reluctant to create a potent central government that might have more fully unified the new nation. Indeed, they were careful to circumscribe severely the competence of the politicians at Philadelphia: the Articles of Confederation drawn up in 1777 reserved to the individual states all powers that were not expressly granted to Congress. But the experience of the war years – and not least the difficulties caused by what Nathanael Greene called 'The local policy of almost all the States'[45] – played a vital part in preparing the ground for a new constitution – a constitution, which, as we will see in the final chapter, attempted to remedy the defects in the Articles of Confederation.

*

The Federal Constitution was a postwar development, but what was the immediate political impact of the war? Under this head we ought to consider a range of topics. The emergence in some of the states of serious political divisions in the revolutionary camp is one that springs to mind. Issues like debt repayment, taxation, currency and the disposal of loyalist property certainly provoked disagreements; but there is no consensus among historians as to the extent to which parties were discernible and how the rival groups should be categorized. Some still follow the class interpretation of the Progressive school of the first half of the twentieth century and see a division between western small farmers and urban artisans on the one hand (the radicals) and wealthy and commercially-oriented merchants and capitalistic landowners on the other (the conservatives).[46] Recently, however, emphasis has tended to be placed on a different fault line, that dividing localists from 'cosmopolitans' or, as some would say, 'nationalists' – those whose connections extended beyond their own states and who adopted a wider world-view; those who, in the words of the North Carolinian William Davie, sought 'to give strength and permanency to the *common union*'.[47]

Divisions within Congress and the Continental army might also be discussed. Congressmen and soldiers disagreed over whether Gates or Schuyler should command the northern army; the rift lasted nearly three years until a new and potentially more damaging dispute surfaced. It used to be thought that there was a serious threat to Washington's position as commander-in-chief in late 1777 and early 1778, when a plot, directed by Thomas Conway, an Irish-born French officer in American service, was supposed to be afoot to install Gates in Washington's place. Gates was certainly popular after Saratoga, and much criticism was directed at Washington for his comparative lack of success in the Pennsylvania campaign of 1777: 'we want you in different places .... We want you most near Germantown', James Lovell told Gates that November. In reality, however, Washington was probably never in much danger; his character, as one of his supporters commented, was 'too well established to be easily injured'.[48] And the number of Washington's avowed enemies was small: indeed, the so-called 'Conway Cabal' is now usually dismissed as the murmurings of a few discontented officers and politicians.

If comprehensiveness were the goal, we should likewise give coverage to relations between the states; relations between the states and Congress; the drawing up of the Articles of Confederation; the formation of the state constitutions. All of these deserve treatment, since all of them were influenced – directly or indirectly – by the war; but in the interests of manageability we shall concentrate on just one major political development, albeit one that can be studied on a number of different levels. The subject in question is the effect of the war on American allegiances; or, to put it another way, the extent to which the conflict changed Americans from discontented but more-or-less loyal subjects of George III into more-or-less willing citizens of the United States.

With the outbreak of fighting in April 1775, it was by no means certain that the colonies would opt for independence. Indeed, a great many Americans clung to the notion that military resistance would oblige Lord North's government to back down; or, better still, would force the administration from office and so pave the way to a negotiated settlement. Hence the use of the terms 'the Ministerial army'[49] and 'The regulars'[50] to describe the British redcoats in 1775. Many Americans – perhaps most – were still genuinely loyal to the crown; it was North and his colleagues for whom they reserved their expressions of hostility. 'O that the wicked, cursed Ministry wou'd but relent 'err it is too late', wrote William Reynolds of Virginia in May 1775, 'and save a number of poor innocent People who must unhappily fall Victim to the designs of a few Men who are a disgrace to human nature.'[51] As Richard Montgomery, the American general who had been in the British army, explained to an old regimental colleague the following July: 'Every Friend to the Empire wishes to throw the odium on ministry, lest the people become implacable against the nation & averse from an accommodation on any terms'. If only the British, he went on, could learn from the colonists 'to resist a wicked ministry – leaving *Majesty sacred.*'[52]

But as the war intensified, and the king made clear his determination to support his government in crushing the American revolt, even the pull of loyalty to the crown began to weaken. The American Prohibitory Act, which aimed to cut off all British trade with the rebellious colonies, declared the Americans to be beyond the king's protection. The ships of the colonists were now fair game for the Royal Navy. John Adams of Massachusetts astutely observed that the British had in effect proclaimed American independence. When it was learned in America that the British were bringing over German mercenaries, the number of advocates of reconciliation dwindled even further. There were still moderate voices that urged caution. Doubting the strength of the United Colonies, they anticipated that the Americans would have to look to France for help. This, they argued, would merely be to exchange one form of dependence for another. But at congressional and state levels the resistance of these faint-hearts was overcome in the spring and early summer of 1776. Thomas Paine's pamphlet *Common Sense* seems to have had a dramatic effect, undermining residual affection for the king and lingering faith in the monarchical system. In July independence was declared.

Independence, however, did not create a united people. Far from it. Independence gave a great boost to loyalism, for it forced many moderate patriots, who could not accept the desirability of the final break, to defect to the British. And there were still many Americans who were uncommitted. Most of the Quakers were principled neutrals. Other Americans were apathetic; or at any rate regarded their mundane concerns as more important than the issue of the Anglo-American contest. Still more, similarly detached from the struggle, adopted a prudential line, trying to steer clear of trouble. James Lloyd II probably expressed the sentiments of many of his countrymen when he told his brother that he looked on 'the present dispute between great Britain & her Collonies, in a very Serious light, & am Determin'd if Possable to avoid taking an active part on either Side'.[53] Nor should we forget those who suffered from the ability to see both points of view – the Americans who opposed British coercion yet continued to feel very uneasy about independence.

Even a protracted war failed to change the political outlook of a good number of these people. There were still many loyalists at the end of the conflict; indeed, between 60,000 and 80,000 of them were sufficiently disaffected to emigrate during or after the war. Silvester Gardiner of Massachusetts, for example, left Boston proclaiming his support for 'the Mild, indulgent Government (God bless it) of England', and, although he found life in Britain rather disillusioning and was bitter about the failure of General Howe to bring the war to a successful conclusion, he continued to abuse and distrust 'the Rebels': in 1782 he even recommended to the British government an assault on Boston.[54] Equally, some of those who were neutrally-inclined in 1775–6 seem to have been no more committed in 1783. Deeply religious Americans, in particular, accepted their sufferings with remarkable stoicism, interpreting the struggle as a divine test or as God-given punishment. Thomas Lightfoot, a Philadelphia Quaker, declared that 'Thro' the permission of Divine Providence there is great desolation in this land', and he concluded that all the ills associated with the fighting 'loudly proclaim the necessity of learning Righteousness'.[55]

We can see, however, that for many Americans the experience of eight years' struggle did leave its political mark. Take Robert Beverley, a Viginia planter of conservative views. He prayed fervently for peace and reconciliation in 1775. In 1782 he still felt uncomfortable with the new regime ('speculative opinions will never do in a *free* country', he warned his son); but he was reconciled to its existence and no longer had any time for the British. Cornwallis's capitulation at Yorktown was described by Beverley as 'a handsome check', and he berated his country's opponents for their 'pride' and 'obstinacy'.[56] When the war began, Evert Byvanck Jr., an elderly New York merchant, was already more well-disposed than Beverley to the Revolution; his views became much firmer, however, during the course of the conflict. At first, despite his patriotic leanings, Byvanck was friendly with John Montresor, a British officer, and Lawrence Kortright, a prominent New York loyalist. His main aim, it seems, was to protect his property. Yet by January 1778 he had been transformed into an enthusiatic revolutionary who execrated the British as 'our Cruel Enemies' and proclaimed that he would not be satisfied with 'a patcht up peace'. A compromise settlement was no longer acceptable; nothing but a total victory would do.[57] A third and final case illustrates a similar process of change this time in a loyally-inclined American. Robert Morton of Philadelphia was the stepson of James Pemberton, a noted Quaker neutral. Morton's own political sentiments seem at first to have been decidedly pro-British. When the British army entered Philadelphia in September 1777 Morton commented that the inhabitants were overjoyed as they had 'too long suffered the yoke of arbitrary Power'. Less than three months later Morton was describing the same British troops in rather different terms. The soldiers he had welcomed as liberators were now 'the Enemy'.[58]

This kind of change was happening not just in a few isolated instances; as the war progressed whole areas seemed to display a greater enthusiasm for the revolutionary cause, or at least more hostility towards the British. True, the opposite tendency can be observed in some places. War-weariness reduced Virginia to near-apathy in the face of successive British invasions between 1779 and 1781. But, as we have seen, in New Jersey, where the local people offered little opposition to the first British incursions, resistance stiffened during the winter of 1776–7: 'the very Quakers declare for taking up arms', a militia officer from Delaware reported to his father on 30 December.[59] The following spring and summer, when the British again tried to advance deep into the state, they met with a fierce reception. The people, according to one account, 'turned out, young and old, great and small, rich and poor. Scarcely a man that could carry a musket was left at home'.[60]

In other areas, resistance took a different form. King's and Queen's counties, on Long Island, New York, were occupied by the British from August 1776. The inhabitants, many of whom were descendants of the original Dutch settlers, were moderate in their political views and keen to live harmoniously with their new masters. An American officer living as a prisoner on parole at Flatbush characterized them as 'people who seemed thoroughly disposed to submit to any power that might be imposed on them'.[61] By the closing years of the war, however, the locals appear to have been engaged in a concerted policy of non-co-operation with the British. There was no armed uprising – in 1782 a British officer could still describe the area around Flushing as 'a Paradise'[62] – but the occupiers found it more and more difficult to requisition their requirements. In July 1780 General James Robertson was 'Mortifyd' to discover that his orders for the supply of goods for the engineers had been ignored by the inhabitants, despite his elaborate efforts to ensure 'that this burthen might be light by being equally born by all'.[63] In November 1781 Commissary Peter Paumier found himself unable to procure wagons from the inhabitants of Queen's County to transport provisions from Flushing to Jamaica township. Around the same time a timber requisition in King's County met with similar lack of success. A 'List of the Inhabitants Who has not furnished the Quantity of Wood' in the commissariat papers reveals that only six of 157 quotas had been met in full, while nothing at all had been provided by nearly half those requsitioned.[64]

So what brought about such changes of sentiment? The answer, of course, varied from individual to individual and from area to area. No doubt prudential considerations were important in many cases: a good number of people probably learned to detect which way the wind was blowing. But there were, in John Shy's words, two institutions which 'played the role of political teacher' in this process of 'political education': the revolutionary militia and the British army.[65] The militia, captured by the radicals right at the beginning of the crisis, performed the function of a political police in many of the areas outside British control: it intimidated and coerced loyalists and moderates, preventing the formation of a viable alternative to the revolutionary regime. But the militia was important in another sense. Nearly every American male, as we have seen, was obliged to make himself available as a militiamen. Continued neutrality was no longer an option when a British attack – and most states experienced them at one time or another – required a militia response. Those who were called out to repel the British were faced with a stark choice. They could refuse the call, be regarded as loyalists, and face the consequences, which were bound to be unpleasant; or they could take the line of least resistance and join their more enthusiastic neighbours. And once mobilized, whatever their original predilections, they were almost certain to be affected by the ethos of this revolutionary institution. To put it crudely, those who were being shot at soon came to recognize their assailants as the enemy and their comrades in arms as friends. It was but a short step from this battlefield assessment to a positive identification with the revolutionary cause.

The British army was important, then, in that its raids forced painful decisions on countless Americans and pushed many of them into the revolutionary camp. But it was important in other ways, too. The army's movements – arriving in an area, appearing to stay, and then retreating – eventually persuaded many Americans who were well disposed towards the British to be cautious about declaring themselves. As a German officer observed at the end of 1780: 'the circumstance that our troops have several times had to abandon places after most of the inhabitants had come out openly for the King, has rendered many unhappy and the rest shy'.[66] Sometimes the effect was more profound: loyalists became disillusioned and convinced that the army could not offer proper protection; once this idea took root their feelings of loyalty diminished.

Worse still, perhaps, both in those areas that the army occupied – like King's and Queen's Counties – and in those areas that it visited more briefly, British troops did much to alienate the affections of moderate and even loyal Americans. The presence of the army did not, admittedly, always have this effect. Newspapers like Rivington's *New York Gazette* and *The Royal Gazette* of Charleston recorded the marriage of several American women to British and German officers, and in these cases we can reasonably assume that the result was the strengthening rather than the weakening of the loyalism of the women's families. And there were clearly opportunities for material advancement while the army was in one's midst. We have seen that merchants, shopkeepers and producers in the occupied towns could benefit from the purchasing power of the military departments and individual officers and soldiers. Yet the presence of the army was not an unalloyed advantage for business: traders complained of British price controls, restrictions on the movement of goods imposed by army and navy commanders ('Many of our Merchants ... have involved themselves in difficulties by vesting considerable Sums in Articles which cannot be shipt')[67] and the poor record of the army in paying its bills ('having ... supplied many Officers of the Garrison with Goods he suffered severely by many of them refusing to pay Accounts against them').[68]

Other features of the army's presence could not even be described as a mixed blessing. The indiscipline of the British troops, we have already noted, only made more difficult the task of winning American hearts and minds: 'no wonder if the Country people refuse to join us', Lieutenant-Colonel Stephen Kemble wrote ruefully in his journal.[69] Requisitioning, the authorized destruction of property, and the encouragement given to slaves to flee from their masters had a similarly damaging effect. So, too, did the billeting of troops in American homes: Joshua Pell complained bitterly that the 27 soldiers in his New York house were 'very Noisy and troublesome to himself and Family'.[70] The army was even blamed for spreading disease; the revolutionary authorities did their best to persuade their countrymen that the outbreaks of dysentery and smallpox in New England in 1775–6 were part of a diabolical plot hatched by Gage and Howe. The failure to restore civilian government in the occupied areas was another grievance. 'There is something in the air of a military government extremely disagreeable to those who have experienced another', wrote William Rawle after living through the British occupation of Philadelphia.[71] Inhabitants of New York City petitioned for a return to civil authority soon after the British army took the town, and as early as January 1777 the Revd. Charles Inglis was expressing regret that they had received no positive response. In Charleston there was a similar clamour for the re-establishment of civil government 'as Soon As the Situation of Public Affairs will permit'.[72] Yet, apart from in Georgia, military rule continued throughout the war. British commanders were understandably reluctant to surrender local control; the revival of civil government, in the opinion of one of Clinton's advisors, could not be accomplished 'without retarding Military Operations'.[73] Even so, such a restoration would probably have had an important psychological effect, undermining the revoltionary claim that the British were determined to establish a tyrannical regime reliant on their bayonets. In any event, the failure to reintroduce civilian government disillusioned a lot of Americans who might have been happy to see a return to the pre-war dispensation.

But perhaps the most fundamental problem was the cultural gulf that separated the army from many Americans. It was as though they inhabited different worlds. Writing from rural New Jersey, Captain the Hon. Thomas Stanley of the Sixteenth Light Dragoons poured scorn on the Quakers who '*thou & thee* me from morning till night';[74] and on Long Island Lieutenant Allan Macdonald of the less fashionable Seventy-sixth Foot was no less disdainful about having to 'reside in a Dutch farm house for near six months in the Year'.[75] Even ordinary soldiers appear to have regarded Americans with contempt: British order-books suggest that commanding officers frequently found it necessary to ask their men to treat loyalists in arms with greater respect. And if to the British the inhabitants were dull, unsophisticated or lacking in polish, to many Americans the army was an agent of moral subversion. Diaries and letters reveal stern disapproval of British drunkenness, gambling and womanizing. Margaret Morris of New Jersey was shocked to see inebriated Hessian soldiers in her New Jersey street and decried their Christmas festivities ('how unlike Christians is manner in which they celebrate it');[76] the extravagance of the farewell for General Howe in May 1778 outraged the Quakers of Philadelphia ('How insensible do these people appear,' wrote Elizabeth Drinker, 'while our land is so greatly desolated');[77] while in New York City the Moravian Pastor Ewald Gustav Schaukirk bemoaned the inadvertant desecration of the graveyard at Trinity Church during the celebration of the British capture of Charleston: 'Profaneness and Wickedness prevaileth – Lord have mercy!'[78] The Revd. Andrew Eliot, who in June 1775 still hoped for a peaceful resolution of the dispute between Britain and the colonies, became a convinced supporter of independence after having lived in close proximity to the army through the siege of Boston. Decrying the behaviour of the British troops – 'Such conduct would disgrace barbarians' – and particularly 'the contempt thrown on our places of Worship', he announced emphatically that 'nothing would induce me again to spend Eleven months in a garrison Town'.[79]

Small wonder that some of the more thoughtful members of the army began to conclude that the British forces had done more harm than good in America. As early as September 1778 Lieutenant-Colonel Charles

Stuart was lamenting that 'We planted an irrevocable hatred wherever we went, which neither time nor measures will be able to eradicate'.[80] Looking back once the fighting had finished, Surgeon John Hayes was equally gloomy: 'We have rambled thro' a Country to make enemies of those, who in their Hearts were friends'.[81]

Chapter 7 Notes

1. *American Campaigns*, i. 249–50.
2. Lord Dunglass to the Earl of Home, 29 Jan. [1778], Douglas-Home Muniments, box 187, bundle 5, The Hirsel.
3. 'News of the Yorktown Campaign', ed. Richard K. MacMaster, *Virginia Magazine of History and Biography*, lxxix (1971), 401–2.
4. Thomas to Sylvester Dering, 11 March 1779, Dering Family Papers, vol. I, William L. Clements Library.
5. *Memoir of General Graham*, ed. James J. Graham (Edinburgh, 1862), p.36.
6. See Julian Gwyn, 'The Impact of British Military Spending on the Colonial American Money Markets, 1760–1783', Canadian Historical Association, *Historical Papers / Communications historiques* (1980), 77–99.
7. Entries of 2, 11, 15, 29 March 1777, Day-book 1774–88, entries of 17 Jan. 1777, 12–23 March 1779, Account-book 1773–90, Faulkner Papers, New-York Historical Society.
8. Clifford to Thomas Franks, 27 Dec. 1777, Clifford Papers, vol. 5, Historical Society of Pennsylvania.
9. Simpson to Germain, 25 Sept. 1780, *DAR*, xviii. 177.
10. *Beekman Mercantile Papers*, ed. Philip L. White (3 vols., New York, 1956), iii. 1304.
11. Lindsay of Dowhill Muniments, GD 254/1055, SRO.
12. Taylor to Neil Jamieson, 2 April 1783, Jamieson Papers, Library of Congress.
13. Chester County Misc. Papers, Depredation Claims, Historical Society of Pennsylvania.
14. *The Papers of Josiah Bartlett*, ed. Frank C. Mevers (Hanover, N.H.: for the New Hampshire Historical Society, 1979), p.195.
15. Wemyss to Cornwallis, 20 Sept. 1780, Cornwallis Papers, PRO 30/11/64, fo. 92.
16. Haldimand Papers, Add. MS 21,760, fo. 32.
17. Loyalist Claims Commission, Audit Office Papers, AO 12/21, fos. 27–8, PRO.
18. Pleasants to Samuel Pleasants, 9 July 1781, Robert Pleasants Letter-book, Library of Congress.
19. *The Papers of James Iredell*, ed. Don Higginbotham (2 vols., Raleigh, N.C., 1976), ii. 328.

20. Samuel Seabury to Myles Cooper, 9 Feb. 1777, Fettercairn Papers, box 75, National Libarary of Scotland.
21. Entry of 16 Oct. 1777, Hasbrouk's 'Book of Record', Library of Congress.
22. Lloyd to ..., 20 April 1783, Lloyd letters, D 1628, bundle 2, Gloucestershire RO.
23. 'Ebenezer Hazard's Diary', ed. Frank Shelley, *New Jersey History*, xc (1972), 172.
24. 'Extracts from the Letter-books of Lieutenant Enos Reeves', ed. John B. Reeves, *Pennsylvania Magazine of History and Biography*, xx (1896), 470.
25. 'Abstract of a letter received from M^r^. Gervais', 27 June 1775, Clinton Papers, William L. Clements Library.
26. Beekman to Hugh Gaine, 23 Dec. 1776, Beekman Papers, box 17, folder 8, New-York Historical Society.
27. Ambler to Mildred Dudley, n.d., but 1781, Elizabeth Jaquelin (Ambler) Carrington Papers, Library of Congress.
28. '"A Diary of Trifling Occurances"', ed. N.B. Wainwright, *Pennsylvania Magazine of History and Biography*, lxxxii (1958), 440.
29. *Letters of Eliza Wilkinson*, ed. Caroline Gilman (New York, 1839), p.12.
30. 'The Cornelia (Bell) Paterson Letters', ed. J. L. Boggs, New Jersey Historical Society, *Proceedings*, xv (1930), 511.
31. Major Israel Keith to Cyrus Keith, 4 Jan. 1777, Israel Keith Papers, New York State Library.
32. *Jefferson Papers*, ii. 3.
33. Phineas Pemberton to James Pemberton, 18 Nov. 1777, Pemberton Papers, vol. 31, Historical Society of Pennsylvania.
34. Quoted in Mary Beth Norton, *Liberty's Daughters* (Boston, 1980), pp.215, 222.
35. Dennis P. Ryan, 'Landholding, Opportunity, and Mobility in Revolutionary New Jersey', *William & Mary Quarterly*, 3rd series, xxxvi (1979), 571–92.
36. See Kenneth Lockridge, 'Social Change and the Meaning of the American Revolution', *Journal of Social History*, vi (1972–3), 403–39; and Rowland Berthoff and John M. Murrin, 'Feudalism, Communalism, and the Yeoman Freeholder: The American Revolution Considered as a Social Accident', *Essays on the American Revolution*, ed. Stephen G. Kurtz and James H. Hutson (Chapel Hill, N.C., 1973), pp.256–88.
37. Gordon S. Wood, *The Radicalism of the American Revolution* (New York, 1992), pp.6–7, 248.
38. *Diary and Correspondence of Count Axel Fersen*, trans. Katherine Prescott Wormeley (London, 1902), p.39.
39. *American Campaigns*, i. 48.
40. 'Diary of James Allen', *Pennsylvania Magazine of History and Biography*, ix (1885), 196.
41. *Letters of Benjamin Rush*, ed. L.H. Butterfield (2 vols., Philadelphia, 1951), i. 244.
42. Information extracted from tax lists and militia rolls for Cumberland County in *Pennsylvania Archives*, 3rd series, xx (1897), 45–9, 166–70, xxiii (1897), 657–9.

43. Elizabeth Hadwen to John Bragg, 12 Feb. 1775, John Bragg Collection, Whitehaven Library, Cumbria.
44. Entry of 11 Sept. 1777, Diary of Richard Bernard, Historical Society of Pennsylvania.
45. *Greene Papers*, iii. 223.
46. See, for instance, Edward Countryman, *The American Revolution* (London, 1987).
47. Quoted in Don Higginbotham, *War and Society in Revolutionary America* (Columbia, S.C., 1988), p.77.
48. *Letters of Delegates to Congress 1774–1789*, ed. Paul H. Smith (21 vols. to date, New York, 1976–), viii. 329, 649.
49. Alexander McDougall to John Jay, 26 Nov. 1775, *John Jay: The Making of a Revolutionary*, ed. Richard B. Morris (New York, 1975), p.181.
50. 'Journal of Ensign Nathaniel Morgan', Connecticut Historical Society, *Collections*, vii (1899), 105.
51. Reynolds to George Flowerdew Norton, 25 May 1775, William Reynolds Letter-book, Library of Congress.
52. Montgomery to Capt. Perkins Magra, 2 July 1775, Misc. MSS (Unbound), William L. Clements Library.
53. 'Papers of the Lloyd Family', New-York Historical Society, *Collections*, lx (1927), 756.
54. Gardiner to ..., 9 May 1776, Gardiner, Whipple and Allen Papers, Massachusetts Historical Society.
55. Lightfoot to Samuel Neale, 30 Oct. 1777, Misc. MSS, Port. 34.180, Society of Friends Library, London.
56. Beverley to William Beverley, 22 Feb. 1782, Robert Beverley Letter-book, Library of Congress.
57. Byvanck to John Byvanck and Garret Abeel, 20 Jan. 1778, Evert Byvanck Papers, New York State Library.
58. 'The Diary of Robert Morton', *Pennsylvania Magazine of History and Biography*, i (1877), 7, 36.
59. *Letters to and from Caesar Rodney*, ed. G.H. Ryden (Philadelphia: for the Historical Society of Delaware, 1931), p.152.
60. 'Diary of Joseph Clark', New Jersey Historical Society, *Proceedings*, vii (1853–5), 96.
61. Quoted in Henry Onderdonck Jr., *Revolutionary Incidents of Suffolk and King's Counties* (New York, 1849), p.173.
62. Journal of Richard Augustus Wyvill, p.16, Peter Force Collection, Library of Congress.
63. Robertson to James Axtell and Archibald Hamilton, 25 July 1780, Henderson of Fordell Muniments, GD 172, box 46, bundle Ra, SRO.
64. WO 60/32, pt. I.
65. John Shy, *A People Numerous and Armed* (New York, 1976), p.216.
66. *Letters from America ...being Letters of Brunswick, Hessian, and Waldeck Officers*, trans. and ed. Ray W. Pettengill (Boston and New York, 1924), p.231.
67. Tench Coxe to Robert Wigram, 23 March 1778, Letter-book, 1776–9, Coxe Papers, Historical Society of Pennsylvania.

68. Petition of Thomas Hopper of Charleston, Loyalist Claims Commission, Audit Office Papers, AO 12/51, fo. 100.
69. 'The Kemble Papers', New-York Historical Society, *Collections*, xvi (1883), 96.
70. Capt. S.P. Adye to Lt. [John?] McNabb, 8 Feb. 1780, Letter-book of Maj.-Gen. James Pattison, vol. ii, Mil Doc 963, Royal Artillery Institution.
71. T.I. Wharton, 'A Memoir of William Rawle', Historical Society of Pennsylvania, *Memoirs*, iv, pt. i (1840), 43.
72. 5 Feb. 1781, Clinton Papers, William L. Clements Library.
73. Memorandum Book, p.51, Andrew Elliot Papers, box 7, New York State Library.
74. Stanley to ..., 23 Dec. 1776, Stanley Letters, MS Am 228.4, Boston Public Library.
75. Macdonald to his mother, n.d. [but Jan.–Feb. 1782], Macdonald Letter-book, Bell-Macdonald Papers, Rammerscales House.
76. Margaret Morris, *Private Journal* (Philadelphia, 1836), p.20.
77. *The Diary of Elizabeth Drinker*, ed. Elaine Forman Crane (3 vols., Boston, 1991), i. 306.
78. 'Occupation of New York City by the British', *Pennsylvania Magazine of History and Biography*, x (1886), 430.
79. Eliot to Isaac Smith, 9 April 1776, Misc. MSS, Massachusetts Historical Society.
80. *A Prime Minister and His Son*, ed. E. Stuart-Wortley (London, 1925), p.132.
81. Hayes to Charles Mellish, 20 March 1782, Mellish of Hodsock MSS, Me 172–111/36b, Nottingham University Library.

8

The Impact of the War on the British Isles

If one reads some of the surviving diaries and letters of the time, one could be forgiven for thinking that for Britain the American war was indeed a 'limited' conflict. The journals of the Rev. Gilbert White, author of *The Natural History of Selbourne*, suggest that the struggle impinged little on the texture of life in his corner of rural Hampshire. Amongst the daily records of weather conditions and observations on flora and fauna, there is the occasional terse reference to wartime events – Rodney's defeat of the Spanish fleet in January 1780; Cornwallis's victory over Greene at Guilford Court House; the 'bloody & obstinate engagement' between the British and Dutch off Dogger Bank; Cornwallis's surrender at Yorktown; naval operations in the West Indies; the siege of Gibraltar. Only once, however, does the war seem near at hand: when in August 1780 White sees Spanish prisoners at Winton: 'rather small men, & some very swarthy'.[1] The records of the less famous can give the same impression. Thomas Falconer, a friend and relative of Charles Gray, the elderly MP for Colchester, regularly commented on the war, censuring the New Englanders in 1775 for 'taking aim at officers with rifled muskets'.[2] But one is left with the feeling that this is just gossip; that it relates to events remote from his personal experience. At no point do his letters suggest that the conflict deeply affected Falconer himself. The Welsh, according to an American Quaker visitor, were hardly interested in the war at all, even as something happening elsewhere: 'they know as little as they care', commented Jabez Maud Fisher in July 1776.[3] Jonathan Adams, a baptist minister living in Salisbury, seems to have been similarly detached. Judging by his diary, he accorded the war almost no importance: it interrupted his crudely written lists of births, marriages and deaths but once, when on receiving news of the battle of the Saintes 'we illuminated [our windows] at night on acc^{t}. of Rodneys success'.[4]

Yet the American war's impact was both wide-ranging and profound.

It was not just a far-off conflict that touched the lives only of those men fighting abroad in the armed forces. Other diaries and letters – and other sources – reveal its intrusive nature. For trading concerns like Alexander Houston and Co. of Glasgow, which imported tropical produce from the West Indies, the American conflict threatened disaster: the firm watched military developments across the Atlantic with intense interest, and prayed fervently for a British victory that would remove the 'great risk and uncertainty' which was having such a damaging effect on commercial operations.[5] And for many more Britons the war was not just something to be followed from a distance. It hardly seemed remote to the midlands clergyman Joseph Greene as he struggled 'to secure my faithful servant from being torn from me as a Militia-Man';[6] nor to Norfolk farmer's wife Mary Hardy when she wrote of her husband's efforts to save a friend from militia service; nor to Richard Neville when his son, a captain in the Berkshire militia, 'Sallied forth ... to attend his Duty in the Field at Reading' and went on from there to spend many months in the encampments established in southern England;[7] nor to the publicans of Lichfield in Staffordshire, who complained of the large numbers of troops and French prisoners whom they were obliged to accommodate. The war must have seemed still less remote to the people of Greenock on the Clyde, where shore batteries were constructed in 1778 to protect the town against anticipated attack; and to Britons all along the south coast as they peered anxiously out to sea waiting for a Franco-Spanish invasion in the summer of 1779. For the sake of order and convenience let us explore the war's immediate impression on Britain and Ireland under the three heads used in the last chapter: economic, social and political.

*

Britain itself was not the scene of military operations, so it avoided large-scale destruction of land-based property. True, we have seen that Jersey was briefly attacked by the French in 1779 and1781 and that the American privateer John Paul Jones raided the coast. We might also note the ripple of alarm in 1776 and 1777 caused by the activities of the pro-American arsonist 'John the Painter', who burnt the ropeworks at Plymouth dockyard and some warehouses at Bristol. But these were essentially minor irritations. Nor, of course, did the invasion that appeared imminent in the summer of 1779 actually materialize.

None the less, Britain did not escape unscathed. Even supposedly friendly armies can make an impression. Scattered through the papers of the secretary at war are letters from civilians protesting at the depredations commited by British soldiers on the march or in camp. In April 1779, for instance, William Jolliffe, MP for Petersfield in Hampshire, complained that the detachment of Guards in his neighbourhood had 'stolen more than twenty Sheep and Lambs a great number of Hogs, and poultry without number';[8] while the following September a militia officer encamped in Kent noted that the local farmers 'have for some time complained of a great deal of mischief being done them by the soldiers, I am afraid with too much truth'.[9] Southern England in particular experienced considerable military activity from 1778, when regular troops and militia were assembled to meet the expected invasion. After his farm had been used as an encampment in May of that year, Henry Penton of Winchester computed his losses – through fences destroyed, crops trampled and so on – at more than £862. Protests were also made about the failure of army officers to settle their bills, and there was a chorus of complaint about the inflationary effects on local prices of the presence of large numbers of military personnel.

Many more Britons were affected by the increased taxation required to fund the war. The land tax, as we have seen, was increased to its normal eighteenth-century wartime rate of four shillings in the pound. Lord North intended that the comfortably-off should contribute in other ways as well: Matthew Flinders, a Lincolnshire surgeon, was pleased to escape the 'new heavy Tax on Male Servants', but he could only reduce his liability for the window tax by blocking up two of the windows in his house.[10] The poor were soon required to dig deeper into their pockets, too, as excise duties were extended to new items of consumption, and the rates increased on goods already subject to duties. The enormous cost of the war was therefore brought home to many ordinary Britons: and the cost was enormous – in excess of £12 million a year on average, and up to a peak of more than £20 million in 1782. Increased taxation, besides provoking screams of protest and much discontent, had a noticeable impact on economic activity. The construction trade, for instance, was experiencing something of a boom until it was badly hit in 1777 by increased duties on glass and wallpaper; bankruptcies in this sector rose from an average of 15 per year between 1772 and 1777, to 30 in 1777 itself and 58 in 1778.

As we saw earlier, government borrowing also increased. This too had a noticeable effect on the economy. To attract savings, the government offered interest rates over the 5% maximum imposed by law on private borrowers. Unable to compete, bankers saw their stocks steadily decline, and were obliged to reduce their new lendings. As a contemporary pamphleteer observed, 'the quantity of circulating cash in the hands of our merchants, manufacturers, builders, improvers, is remarkably diminished.'[11] The overall pattern was clear: the government's need for money was deflecting investment from building, from canal and road construction, and from the land – the war years coincided with a sharp decline in enclosure activity.

The problem was exacerbated by the disruption of overseas trade, which lowered profits and so further depressed investment in the domestic private sector. Smugglers no doubt flourished with the navy's attention distracted, but legitimate trade declined: between 1775 and 1778 imports fell by 26%, exports by more than 18%, and re-exports by 35%. The contraction owed much to the near cessation of trade with America. In July 1775 the Virginia agent of a Glasgow tobacco merchant predicted that the war would have a 'direful effect' on business.[12] He was right: Scottish tobacco imports slumped from 46 million pounds weight in 1775 to only seven million in 1776. By 1777 they were languishing at a mere 210,000 pounds. Of course, for the bigger firms, which had imported large volumes of tobacco in anticipation of the crisis, the fall of imports was a blessing – at least for a while. High prices in 1776 and 1777 increased profits. As late as August 1777 Glasgow merchants were reported to have 'sold a great quantity of tobacco to the French at a very great price'.[13] But once the stocks were exhausted there was no replenishment and profits disappeared.

Those who exported to America were also badly hit. Small quantities of goods continued to enter British-held enclaves like New York, and hopes were frequently expressed that British military success would soon make inland markets accessible again. Such hopes proved false for all but the briefest of periods. 'All prospect of the Countrys opening is now at an end for this Campaign', Frederick Rhinelander, a New York merchant, told a British supplier; to another he regretfully explained: 'When we sent you this Order we certainly presumed on a Conclusion of this unhappy Contest and of course an open Country'.[14] The sharp decline in the American trade seems to have been particularly damaging for the textile industry. In 1772 the official value of woollen exports to the 13 colonies was more than £900,000; by 1776 this had fallen to a trifling £58,000. In the West Riding wool producing area, it was not until the close of the war that output consistently returned to the level of 1773. This inevitably had an impact on the wool-growing regions. Wool prices slumped and wool growers, especially in Lincolnshire, became restless. In 1780 and 1781 they petitioned parliament – though unsuccessfully – for permission to export raw wool to Europe. The loss of most of the American market also had damaging consequences for the metal-working trades of the West Midlands. In 1776, Andrew Oliver, an American loyalist refugee, could still wonderously describe Birmingham as 'a perfect Beehive'. By 1779 he was noting the 'Decay' of nearby Walsall.[15] The staunchly pro-American

Sylas Neville recorded in his diary for October 1781 that Birmingham's Soho Works had employed 1,000 operatives before the war, but now had work for only half that number. 'This is a confirmation', he added sarcastically, 'of what we are told by our worthy governors & their tools that Birmingham & Sheffield are not in the least affected by the American troubles.'[16]

But the collapse of the American market was not wholly to blame. Once France and Spain entered the war, many European markets also closed, or at best became less accessible, exacerbating the problems caused by the outbreak of hostilities with America. Indeed, in many sectors of the economy serious difficulties were experienced only from 1778 and 1779 rather than from 1775. Exeter's exports of serge cloth to Spain were worth about £150,000 in 1778. In the next two years they dwindled to almost nothing. Similarly, the city's trade with the Italian states fell by over 70% between 1775 and 1780, most of the decline coming with the interruption of Mediterranean trade once France and Spain became Britain's enemies. The already beleaguered construction industry had to contend with new impediments to the timber trade with Scandinavia; and of course all overseas traders were affected by the increased insurance rates required to cover the risk of sailing in seas where enemy naval vessels and privateers were likely to strike. Irish discontent, which we will consider shortly, added to the gloom. Following American example, the Irish instituted commercial boycotts to coerce the British government. From Bolton in Lancashire, Josiah Wedgwood reported in October 1779 a sharp rise in local unemployment and noted that 'The manufacturers say the measures which the Irish have adopted in their non importation agreements, have affected their trade very much.'[17]

Another factor affecting trade was the navy board's demand for shipping to transport troops and provisions to the overseas spheres of operation. Great numbers of vessels were taken out of mercantile service to meet this requirement. By December 1781 more than 113,000 tons of shipping was under hire to the navy board – the equivalent of about one-sixth of all British merchant tonnage. And of course a further and still more important contribution to trade contraction was the capture by the Americans, French, and Spaniards of an estimated 3,386 merchant vessels in the course of the war.

There was, however, a more positive side as well. If the British merchant marine suffered from the depredations of enemy vessels, British merchant ships acting as privateers made their own captures. In Liverpool alone there were said to be 120 privateers operating in 1779, and the profits they made provided some compensation for the town's wartime commercial losses. Vessels hired out to the navy board and other service departments reduced the number available for normal mercantile purposes, but the owners were able to earn a decent income – at least until the later stages of the war, when the rates offered by the government failed to rise sufficiently to compensate owners for the inflation of their costs. We should note also that the collapse of exports to the 13 colonies was to some extent offset by the increased trade with other areas: indeed, the Americans may well have continued to consume significant quantities of British goods which were imported indirectly either from Europe or the West Indies. The value of Exeter's serge exports to Flanders rose from £58,000 in 1775 to £270,000 in 1780; while during the same period Scotland's exports to the West Indies increased from £71,671 to £256,193. Even the sparcely populated Floridas started to receive increased imports: their trade with Scotland before the war had been negligible; by 1777 they were importing £56,516 worth of Scottish products.

And new opportunities arose within Britain. The military camps established in southern England offered commercial advantages for those willing to seize them. Coxheath Camp near Maidstone was home to 17,000 regulars and militia in 1778, and occupied a frontage of more than three and a half miles. An assembly room was even built, at which weekly concerts were performed by military bands to entertain both soldiers and visitors. And the number of visitors was clearly significant. According to one officer the inns in Maidstone 'from the great resort of people to this camp, ... are generally full'.[18] Warley Camp, near Brentwood in Essex, was smaller, but still drew multitudes of relatives, friends and merely inquisitive onlookers. Money could be made from both resident soldiers and sight-seers; through prostitution, side-shows, the sale of refreshments, and even the organization of regular coach services to take visitors from London to the camp sites: Jeremy Bentham, the utilitarian philosopher and jurisprudent, was one of those who went to satisfy his curiousity, and while travelling along the road to Brentwood he 'overtook a vehicle that called itself Warley Camp Common Stage Cart. It contained 3 benches with 3 people on each bench who sat one behind another ... as in a Playhouse'.[19]

The expansion of the armed forces brought other benefits. The textile industry found some compensation for the loss of the American market in more orders for military uniforms. Sometimes, indeed, this new business was much more profitable than the old. William Wilson and Son, tartan manufacturers of Bannockburn, sold large quantities of plaid, kilts and hose to merchants supplying both the old Forty-second Foot and the newly-raised Scottish corps. Sales of this type far exceeded the value of any non-military orders, and the company found itself so busy catering for the requirements of the army that it was obliged to turn away other customers: 'since we left your place', Alexander Fraser of Forres was told on 9 November 1778, 'we have Engaged in the Cloathing for some of the Highland Reg^{ts} which puts it out of our power to get your order ansrd at this Time'.[20] Naval expansion provided a similar stimulus: the dockyards grew as more ships were built and required repair work. And for the munitions-makers this was naturally a good time. Samuel Walker's Rotherham ironworks were reported to have 1,000 employees in 1777; and the firm doubled its capital between 1774 and 1782, thanks largely to profits from the manufacture of cannons. Less obviously, perhaps, the grain producers of East Anglia and the livestock farmers of Ireland benefited from the demands of the British forces fighting in the 13 colonies. Before the war, the British garrison in America had largely been supplied from colonial sources. Once hostilities commenced, the greatly enlarged British army in America had to rely heavily on provisions shipped 3,000 miles across the Atlantic.

Nor should we overlook the way in which the war improved the lot of at least some working people by increasing their wages. This is not to deny that the opposite could happen. Trade depression meant that in 1778 framework-knitters could make only seven shillings a week – much less than before the war. And, as we have seen, there was certainly unemployment, which usually keeps wages steady or drives them down. But we should note that recruitment into the army, pressing for the navy, and service in the militia was at the same time taking a significant number of men out of the labour market. Military and naval mobilization, moreover, soon went far beyond the ranks of the unemployed. Employers regularly complained of the loss of skilled men. The government was sympathetic; indeed, it was illegal for apprentices to be recruited into the armed forces and those who willingly joined might be subject to punishment: a Worcester apprentice who sought to escape his master was fined and imprisoned 'for a Fraud – in receiving the Bounty Money for inlisting as a Soldier'.[21] Even so, it was not always easy to prevent such enlistments: Samuel Knowles, a Sheffield cutler, was just one of the employers informed that his apprentice had 'embarked for N^{o}. America' before the war office could intervene to secure his discharge.[22] The impact on wage rates varied from region to region and occupation to occupation. Labourers at Maidstone found their daily pay increase from 1s 5½d in 1775 to 1s 7d in 1776 and to 1s 7½d in 1780. It slipped back slightly to 1s 7d in 1781, but stabilized at this rate for the rest of the war. In and around Guildford in Surrey, craftsmens' wages followed a similar pattern: they climbed steadily from 2s 2d in 1774 to 2s 3d in 1776, to 2s 5d in 1777, and reached a wartime peak of 3s in 1778. They remained at this level until 1783, when they fell back to 2s 5d. Of course, such pay rises have to be put in context. They were probably insufficient to keep pace with the cost of living. But there were working people who were more fortunate. Those merchant seamen who avoided the press gang saw a spectacular growth in their monthly pay, which rose from £1 10s in 1776 to £3 5s in 1783.

The picture just painted is highly variegated. Historians disagree about the impact of war on the eighteenth-century economy, and such is the complexity and range of the issues to be weighed that a proper audit is probably impossible. But whether we are focusing on costs or benefits, losers or winers, one thing is clear. The American war made its mark.

*

The social impact of the American conflict is in some respects even more difficult to assess, because the information available to us is very patchy. There may have been some tentative steps forward for British women. Female earnings were vital to the incomes of many poor families long before the war, but it seems likely that more women were employed – and perhaps at different types of work – as men left the labour force for military or naval service. We can only speculate on how relations within families were changed once women acquired, even if just temporarily, new responsibilities. But it seems unlikely that returning soldiers, sailors and militiamen always found that family life automatically reverted to its pre-war pattern. And can we detect, perhaps, a new female assertiveness in the upper-class vogue of 'cross-dressing' – the modification of women's dress to make it more suitable for 'manly' outdoor pursuits like hunting? This development caused especial comment during the war when women of fashion started to wear versions of uniform coats, complete with military-style trimmings.

But such miniscule advances – if, indeed, they were advances – were more than outweighed by other wartime developments that emphasized women's traditional role as dependants and mothers. Cross-dressing was mercilessly pilloried in cartoons: as military success eluded the British, female attempts to encroach on male preserves were increasingly seen as 'unnatural' symptoms of all the ills of the age. And, of course, nothing brought home the subordinate and dependant status of women more forcefully than the death of a male partner. Elizabeth Fulton, mistress for nine years of Captain John Ridley of the Twenty-eighth Foot, was 'left with two helpless Infants, without a possibilitiy of supporting them', when Ridley died at New York in December 1776.[23] The widow of Sergeant Hutchinson of the Eightieth Foot can hardly have been better off when he succumbed to 'a very severe Illness' on the voyage from

Scotland to America in 1779.[24] Moreover, government support for military widows, and even the wives and children of militiamen on service, only served to underline the place of women as home-makers and mothers: it was as if the local or national agencies of the state were acting as surrogate husbands and fathers. In Berkshire, for instance, Martha Day, the wife of Richard Day 'a hired Substitute to Serve in the Room of W^{m}. Whitney', was paid two shillings a week – for at least seventeen months – by the parish of St Nicholas in Abingdon to support herself and their child.[25] Likewise, the respectable and dutifully suppliant Elizabeth Hall, 'left with Four small Children without any Means of Subsistence Whatsoever' when her husband in the Westmoreland militia died in 1779,[26] could realistically hope for some relief when she applied to the secretary at war. By contrast, women who were no less desperate but who deviated from the idealized norm usually received no favour. At a time when large bodies of soldiers were on the march or in camp all over the country, prostitution was seen as particularly dangerous. Ann Griswell and Ann Melow, charged in 1780 by an officer of the East Yorkshire militia 'with being common prostitutes, wandering abroad, infecting the said reg^{t}.',[27] could expect nothing but punishment. In this sense, the American war – like other eighteenth-century wars – reinforced the male stereotypes of women as saints or harlots.

Traditional thinking was bolstered in another area. The rural squires involved in the county protests of 1779–80 seem to have been influenced by hostility to the rise of an allegedly new breed of monied men – 'stock jobbers' and contractors whom the country gentlemen believed were absorbing most of the government's bloated tax revenue and were thereby making Britain's considerable war effort ineffective. This was nothing more than a revival of the anti-commercial rhetoric of the squirearchy in the days of Queen Anne, and it had even less basis in truth than earlier arguments against the *noveaux riches*. Many of those who did well out of the war were in fact themselves from landed families and not part of a discrete 'monied interest'. The Scottish baronet Sir William Erskine, quartermaster-general in America, made substantial sums – often, it seems, by dubious means. To Sir John Halkett he wrote on 14 January 1777: 'by my next letter I shall empower you, to pay off betwixt two or three thousand pound of my debt, this you must not make known to any body'.[28] By the end of the following August Erskine had sent home some £13,000. Nor should we forget that the squires in the shires were not the only critics of profiteering. Samuel Kenrick, a middle-class manufacturer and banker who opposed the war, was no less bitter in his denunciation of contractors who grew fat during the conflict; and Kenrick added a new twist to the time-honoured rhetoric by pointing out that 'a parcel of useless idle beings, called gentlemen's youngest sons' benefited from commissions in the expanded army, and then had to be maintained at great public expense on half pay once the war was over – an interesting anticipation of nineteenth-century attacks on the vested interest of the landed classes in international conflict.[29] Even those few beneficiaries who, with a little imagination, could be portrayed as less elevated 'new men' turn out on closer inspection to defy the steroetype that the country gentlemen had in mind. Anthony Bacon, an army provision contractor and supplier of cannons to the ordnance, was able to retire on his profits in 1782; but he had hardly been an outsider when the war started: he had held government contracts since 1758, he had become a substantial leasor of land in south Wales in 1765, and he had been MP for Aylesbury since 1764. Richard Sadleir, a Southampton brewer and merchant, made enough from his government provision contracts to help establish the town's first bank in 1778, and within a short time he was its dominant partner. But he owed his contracts to his family's local prominence: his father, a wealthy man, enjoyed significant electoral influence in the borough.

Squirearchical apprehensions were of course inextricably connected with a sense of their own loss of power and status. If they believed that there was a rising and socially distinct 'monied class', the country gentlemen tended to assume that they themselves must be a declining force. In some cases, however, social decline could be absolute rather than relative, and real rather than largely imagined. James Hamilton, a young merchant from Lanark, seems to have been so badly hit by the disruption of the Anglo-American trade that he took the desperate step of joining the Third Foot Guards as a private.

For others, military service provided the possibility of social advancement. We can only guess whether Edward Kitson flourished in the postwar years, but the war itself certainly helped him. In August 1776 he was enlisted in Shrewsbury gaol by Lieutenant John Ridout of the Forty-sixth Foot. Kitson had been confined for a petty offence and Ridout was happy to take him into the regiment. Within a year of joining the Forty-sixth in America, Kitson became a corporal. In due course he was promoted to sergeant, and he continued to hold that rank until he was released from service on 4 April 1783 after a period 'On furlough' – a sure sign that he was regarded by his officers as trustworthy.[30] A spell in the army had transformed Kitson, it seems, from a lowly prisoner to a respected veteran. Samuel Bradford, the son of an Irish linen-draper, saw his star rise even more spectacularly. In 1780 he was given an ensigncy in the newly-raised Ninety-second Foot, largely, it appears, on the recommendation of a local gentleman who wanted to encourage recruitment in the area. Young Bradford's good fortune was tragically brief. Later that year his regiment was ordered to the West Indies, where on 25 December he was mustered with the rest of his company at Spanish Town, Jamaica. Thereafter, however, he disappears from the records: it seems likely that, like many of his colleagues, he succumbed to one of the deadly diseases prevalent in the Caribbean.

James Green was luckier. He joined the Sixty-second Foot as a private in 1772, and worked his way up through the non-commissioned officer ranks to sergeant-major. This was as far as he could realistically expect to go. But then in 1777, when his regiment was short of officers, he was made an ensign. Two years later he transferred to the Twenty-sixth Foot as a lieutenant, and drew extra pay by serving as the regimental adjutant. He impressed his superiors so much that he was appointed a deputy judge advocate in America, in which capacity he supervised, according to his own account, 'near fifty Trials'.[31] We can surmise that life was not always easy for Green: for several years the officially-produced *Army List* seemed determined to highlight his humble background by refusing to acknowledge that he had a Christian name. In due course, however, he became indistinguishable – at least in print – from his regimental colleagues, and rose by steps so that by 1803 he was the lieutenant-colonel. Promotion from the ranks was far from unknown in the eighteenth-century – indeed, it occurred frequently enough to provoke adverse comment in some quarters – but in peacetime it was very rare, and we can be confident that Green owed his progress to the American war.

Moving to an entirely different area, we can see that the war also influenced penal policy and the pattern of crime in Britain. Before 1775 it was standard practice for felons to be pardoned on condition that they were transported to the American colonies. Lesser criminals might be sentenced to transportation by the courts. The outbreak of fighting between Britain and the 13 colonies made the continuation of this practice impossible. Once it became clear that the interruption in the export of offenders was to be prolonged, new methods of dealing with the deviant population had to be devised. Male offenders – and even suspects – might be packed off to the army, navy or marines, or, like John Whiteley, found guilty of horse-stealing at the York assizes in 1776, were pardoned on condition that they entered the king's service either 'at Sea or Land'.[32] But the number absorbed by the armed forces was limited; not least, as we saw earlier, because the war office and officers on the spot were often reluctant to take recruits of this type. Many convicts were subjected to a novel form of punishment – hard labour on hulks moored in the Thames. The emphasis thus shifted from disposal of offenders – by execution or transportation – to custodial sentences and reformation: hard labour was intended to improve the character of the prisoner and make him (or her) fit for release.

One should be careful not to exaggerate the extent of the change. Once the war ended, transportation recommenced – at first to the coast of Africa and then, from 1787, to Australia. But the idea of imprisonment was given a great boost by the practice of imprisonment more or less forced on the authorities by the wartime collapse of the transportation system. Detention and forced labour in the hulks added strength to the arguments of those who advocated the redeeming qualities of incarceration and disciplined work – reformers like Jonas Hanway and John Howard, who now found their theories gaining ground. It was as a result of their influence that parliament passed the 1779 Penitentiary Act, which envisaged the creation of a series of penitentiaries – new-style prisons in which the inmates would be subjected to a rigorous regime of hard labour, religious instruction, and solitary confinement at night. Though it was never put into effect, the act encouraged local emulation in the form of the penitentiary-building initiatives of the Duke of Richmond in Sussex and Sir George Onesiphorus Paul in Gloucestershire. If the American war did not transform British penal policy, it certainly gave it a hefty push in a new direction.

Crime levels were probably also affected by the conflict. During the American war, as in previous eighteenth-century wars, the number of prosecutions for offences against property fell; and once the war ended prosecutions sharply increased. One could argue that this is a misleading indicator of the crime rate. We have already seen that soldiers in camp or on the march were frequently accused of all manner of thefts: their crimes would have been punished (if at all) by courts martial rather than by the normal courts. And if civilian offenders were encouraged to join the army or navy prior to indictment, their names would never have appeared in the criminal records either. A fall in the number of prosecutions at quarter sessions and assizes might therefore be no guide to the number of crimes committed: indeed, it might mask a crime rate that remained static or even rose.

But it seems likely that there was a real decline in the crime rate during the war. The recruitment into the armed forces of much of the surplus labour supply removed a good few *potential* miscreants from circulation; and at the same time made things easier for those working people who remained behind, thus reducing the economic pressure to slide into criminality. This interpretation chimes nicely with the steep rise in prosecutions with the return of peace. At the close of the war, thousands of soldiers and sailors were demobilized. Their return to civilian life meant that the labour market became overstocked. It was a commonplace observation in 1783 – as it had been in 1763 and 1748 – that ex-servicemen often had difficulty in finding work and so turned to crime. Some commentators believed that service in the army, navy, or militia was itself part of the problem: it was said to be brutalizing, and a school for crime; or, because of its long spells of inactivity, to destroy the desire for work, making labourers and artisans lazy and unfit to resume their former stations. In 1783 the *Salisbury Journal* reported that 'a great number of disbanded militia-men, who are too idle to return to their farming business, are robbing all parts of the country'.[33]

Service in the armed forces may have had other effects. It probably furthered a sense of national identity. Even for non-combatants in eighteenth-century Britain, wars played a part in shaping the nation, especially wars against France and Spain. France, in particular, as the nearest and most threatening, was the obvious defining counterpoint. 'The natural antipathy I believe we all have towards the French had rous'd the Britain [sic] in my soul', wrote a young Oxford undergraduate in August 1778.[34] The alleged poverty of France was seen as an inevitable consequence of her Catholicism and her unchecked, arbitrary government. In Britain, by contrast, it was assumed that Parliament, a limited monarchy and Protestantism underpinned national prosperity. Cartoons depicted Frenchmen as emaciated and wearing wooden shoes, living on garlic and frogs, or as aristocratic fops. The Englishman was always shown as stout and plain-dressed, a frothing tankard of ale and a substantial joint of roast beef his almost constant companions. John Bull, who first appeared in a pamphlet at the close of the War of Spanish Succession, and was immortalized by James Gilray in the 1790s, finds his way into the prints of the American war as Jack English, a very similar caricature. But for those who actually bore arms, there was a practical dimension that bolstered the propaganda.

Consider the soldiers of the English and Welsh militia. For men who had lived their lives confined to their own locality – and this was probably true of most of them – militia duty brought a dramatic widening of horizons. In the summer of 1779 the troops encamped at Coxheath included militia units from Devon, Suffolk, Monmouthshire, Nothamptonshire, Buckinghamshire, Norfolk, Somerset, Dorset, Gloucestershire, Warwickshire, Yorkshire, Lincolnshire, Rutland, Angelsey and Carnarvon. At the same time Warley camp was home to militia from Berkshire, Cambridgeshire, Hertfordshire, Kent, Middlesex, Cheshire, Derbyshire, Pembrokeshire and Radnorshire. The next year operations against London's Gordon rioters employed militiamen from as far away as West Yorkshire and Northumberland. Returning home at the end of their service, many of these temporary soldiers cannot but have seen the world and themselves in a different light. Their military experiences must surely have made them more self-consciously English, even British.

A sense of Britishness might well have been promoted within the armed forces and among the people at large by significant numbers of Scotsmen serving in a military capacity. Before the conflict, only four of the 70 line infantry regiments of the regular army had strong Scottish associations, though many Scots were to be found in other corps. Of the 35 infantry battalions raised between 1775 and 1781, no fewer than 12 were distinctively Scottish (not counting two battalions of Highland immigrants raised in America and put on the regular establishment). And all this while Scotsmen continued to be over-represented in non-Scottish units. As in the Seven Years' War, conspicuous Scottish military participation helped to neutralize anti-Scots feeling in England. This is not to say that all English hostility was overcome. Far from it. Radicals viewed the Scots as authoritarian, as still infected with Jacobite principles. Before the war, John Wilkes had made his name by attacking the power enjoyed by Scotsmen like Bute, the king's first minister, and William Murray, Lord Mansfield, the lord chief justice. On the eve of hostilities, radicals even suggested that only Scottish troops would be willing to suppress the disaffected colonists; English soldiers were portrayed in a radical cartoon as refusing to fight their fellow countrymen in America. From Boston Earl Percy wrote in early 1776 that Brigadier James Grant's hold over General Howe would convince the English radicals that 'the Scotch influenced the Cabinets here as well as at home'.[35] But Percy was more amused than alarmed by Grant's pretensions, and was poking fun at the Wilkite radicals. In the army the Scots were able to prove themselves to be good Britons – just as good as the English. Indeed, some Scots officers preferred not to serve in specifically Scottish corps, partly, as one confessed, because 'Promotion is slow',[36] but also, it seems, because they wanted to be considered as *British* officers in a *British* army. The result of this willingness to become British, and particularly to serve in such a British institution as the army, was that English memories of the Forty-five Rebellion and Scottish invasion were overlaid by images of Scottish loyalty and service. The staunch defence of Minorca by the Scottish General James Murray, and particularly his refusal to accept a bribe offered by the allied commander, marked him out as a true British hero: 'The prop of Britain's sinking name', according to a contemporary ode, 'The bullwark of her falling fame'. Indeed, Murray was even ranked alongside such seventeenth-century *English* heroes as John Hampden and Algernon Sidney, who had made their libertarian reputations by resisting arbitrary Stuart (and therefore Scottish) monarchs.[37] Despite Wilkes and his supporters, by the end of the war English acceptance of the Scots was all but complete.

A sense of Britishness was no doubt butressed by British art, which itself was influenced by the impact of military service on the national consciousness. Popular fascination with all things military – when the newly embodied Northamptonshire militia were inspected in May 1778, Lord Althorp noted the 'great number of people who notwithstanding the rain had come to see the review'[38] – encouraged artistic endeavour in a variety of areas. Philippe de Loutherbourg painted the manouevres at Warley Common, cartoons revealed the supposed pleasures of Coxheath – the name gave ample scope for sexual allusion – and there was even a musical comedy 'Warley Camp', composed by Charles Dibdin with words by Richard Brinsley Sheridan. Rodney's victory at the Saintes was also exploited to the full by the production of a mass of cheap ceramic souvenirs. At a more rarified level, paintings celebrated other British triumphs. Perhaps the most famous was John Singleton Copley's 'Death of Major Peirson' (1783), which captured in dramatic form the end of one of Britain's war heroes.

*

For a supposedly 'limited' war, the American conflict had some important political repercussions. It was directly responsible for Ireland gaining greater independence. Irish Presbyterians' sympathy for the Americans owed much to the emigration of many Ulstermen to the colonies; the Anglican bishop of Derry went so far as ascribe America's rebellious spirit to the recent arrival there of more than 30,000 fanatical republicans from the north of Ireland. But Protestants in general – Anglicans as well as Presbyterians – tended to be pro-American. This was partly in reaction to the Irish Catholics who seemed keen to show their loyalty to the British – Catholic peers and gentlemen offered bounties to encourage their co-religionists to enlist in the army – but mainly because the parallels between the constitutional situations of the Americans and the Protestant Irish appeared so close. Although Ireland had her own Parliament (in which only Anglicans were supposed to sit and which was elected only by Protestant voters) Poyning's Law of 1494 subjected it to the control of the English Privy Council; and in 1720 the Westminster Parliament asserted its right to legislate for Ireland in all cases whatsoever.

On a practical level, British restrictions on Ireland's overseas trade caused particular resentment. This intensified with the coming of the war, because in February 1776 an embargo was placed on the export of Irish foodstuffs. The aim, it seems, was to ensure that no valuable supplies reached the Americans; but it was widely assumed that the British government was keen to keep the price of Irish provisions stable at a time when it was making significant purchases for the armed forces serving in the colonies. The lord-lieutenant, the Earl of Buckinghamshire, was sympathetic to Irish economic grievances, and he supported the calls of the Irish Parliament for trade concessions. By Early 1778 Lord North's government in London was preparing to overhaul the protectionist system that regulated Ireland's foreign commerce; but howls of protest from British manufacturers, fearful of Irish competition, ensured that the government's plans were considerably diluted. The few concessions made failed to satisfy the Irish, who now adopted non-importation agreeements directed against British goods. But it was the emergence of large bodies of armed 'volunteers' that finally persuaded Lord North to press ahead with the redress of Irish grievances.

The withdrawal of much of the British army garrison to fight in America had brought about the rise of the volunteers. By August 1778 the remaining regulars were so few in number that the commander-in-chief argued that not even two companies could be spared for the defence of the Isle of Man. In the absence of an Irish militia, volunteer units more or less spontaneously emerged to protect the country from Bourbon invasion. Although supplied with arms by the government, these units were not under the control of the British authorities at Dublin Castle. The British and Irish governments must have been especially alarmed when they saw the popularity of volunteering in Ulster, where the Presbyterians had shown a marked reluctance to join the British army to fight against 'their Dissenting bretheren in America'.[39] The ambiguous nature of the volunteers was only too apparent to the lord lieutenant. He chose to adopt a metaphor, but his meaning was perfectly clear: 'the lyon walks abroad without his chain, and tho' he wags his tail, his fangs may prove dangerous'.[40] Irishmen anxious for reform had no doubt that they could take advantage of British unease: William Brownlow believed that the appearance of the volunteers lining the streets of Dublin while the Irish Parliament met 'will be the cause of some reflection on the other side of the water'.[41] It certainly was: in October 1779 Lord North was fearful that Ireland was going the same way as America; while his secretary to the treasury, John Robinson, argued that Ireland would be lost if French money and arms reached the country. It soon emerged that there were Presbyterian volunteers who even seemed prepared to risk Protestant hegemony by advocating the repeal of anti-Catholic legislation, so strong was the new sense of patriotic solidarity and so deep was their conviction that Catholicism could not florish in an atmosphere of true religious liberty. From 1782, when a great volunteer convention at Dungannon pressed for the removal of all legislative restrictions on Irish independence, some of the volunteers adopted a still more radical agenda, calling for parliamentary reform in Ireland itself. That August, a British visitor to Belfast wrote that 'The Original Armers of the people begin to tremble', that the volunteer units 'are dayly dismissing their officers', and, most ominously, that Belfast itself was 'a perfect Boston'.[42]

Within a short time, however, the volunteers began to fade; outside Ulster they rapidly became insignificant. Their period of greatest influence was the war years 1779–82, and their triumphs were limited to reforming relations with Britain. Allied with the opposition in the hitherto emasculated Dublin Parliament, the volunteers forced first commercial concessions from Britain and then constitutional changes. In 1780 Ireland was granted freedom to trade with the British colonies and restrictions on the export of woollen goods were removed. In the same year Protestant dissenters were allowed to take public office by the repeal of the relevant clauses of the Irish Test Act of 1704. In 1782 the British parliament repealed the Irish Declaratory Act and the Irish parliament amended Poyning's Law to remove Privy Council supervision of Irish legislation. In response to further pressure, in 1783 the British government secured the passage through Westminster of a Renunciation Act, which specifically renounced British legislative authority. Ireland had not become a fully independent sovereign state, but it had achieved a degree of legislative independence that makes it far from overfanciful to talk of an Irish Revolution that followed on from, and was made possible by, the revolution and war in America.

In Britain itself, the war was similarly responsible for extra-parliamentary pressure for constitutional reform. Increased taxes, a widespread suspicion that public money was being squandered on buying parliamentary support for the government, and – above all – lack of military success, had by the summer of 1779 created a mood of great discontent. The loss of control of the Channel to the French and Spaniards was perhaps the final straw: as the British fleet remained inactive Sir William Meredith, once a parliamentary supporter of Lord North but now an opposition MP, wrote gloomily of a 'fatal torpor which hangs like the night-mare over all the powers of this Country'.[43] By the end of the year dissatisfaction was beginning to find expression in what what has been described as a movement for 'National Revivial'.[44] Central to this movement was the desire for an end to the undue influence exercised by the executive on the legislature through its control of the voting habits of MPs who were financially dependent upon the government. As the Yorkshire squire Sir James Norcliffe wrote in December 1779, what he wanted to see was 'the constitution so far restored as to have a representation independent of the influence of the executive part of the constitution'.[45]

This executive influence was thought to be growing as the number of dependent MPs increased, and it was believed to be steadily reducing Parliament to the status of a mere tool of the government, a rubber stamp for the decisions of ministers. This subservience, it was widely assumed, explained the poor running of the war – ministerial mistakes remained unchecked because Parliament was losing (or even had lost) its independence. Petitions from nearly 40 associations and committees in counties and boroughs from across the country arrived at Westminster, with the Yorkshire Association, led by Christopher Wyvill, setting the pace. Their common demand was 'economical reform' – a reduction in the influence of the king's ministers by the cutting of the number of places in government pay and by preventing certain office-holders from sitting in the Commons.

Many of the squires in the counties wanted to go no farther than this. Initially, the most ambitious demand of the rural wing of the association movement – and one that some conservative squires were reluctant to embrace – was for the creation of 100 new county seats (abolition of rotten boroughs received some early support and later became part of the programme). But even the county-seat panacea can be viewed as a manifestation of the traditional 'country' concern to counter executive influence and preserve the independence of the House of Commons as a check on the government: the counties had larger electorates than most boroughs and county members most nearly approximated to the ideal of the independent-minded and incorruptible legislator. The radical fringe of the association movement, however, based in the metropolis and particularly in Westminster, went much farther, or rather was moving in a different direction. The Westminster committee pointed out that the large number of men in military service was an argument for a dramatic extension of the franchise: why should a man be asked to risk his life for his country yet 'be thought unworthy of a suffrage in the election of his representative'?[46] In 1780 the committee announced a truly democratic programme – including universal manhood suffrage, the secret ballot and equal electoral districts – a programme that anticipated the demands of the radicals of the 1790s and was to form the creed of the Chartists in the 1830s and 1840s. The French Revolution may have broadened the social base of the British reform movement by bringing in significant numbers of plebian activists, but the ideas they espoused were essentially those of the Westminster radicals of more than a decade before.

Having stimulated a movement for reform and revival, the war then helped to undermine it. 1779 had been a very bad year in military terms for the British. 1780 was different. In May Charleston fell and Clinton and Cornwallis seemed set fair to subdue the southern colonies. This no doubt played a part in dampening down much of the domestic discontent. British success was of course only temporary; but although further calls for reform were made in 1781, 1782 and 1783, they received noticeably less support than the great campaign of 1779–80. Perhaps both the original decline of the reform movement and its failure to regain its former strength should be attributed in large measure to the influence of the Gordon riots of June 1780. This spectacular eruption of anti-Catholicism in London followed on from similarly violent unrest in Scotland in 1779. The activities of the fanatical Lord George Gordon's Protestant Association, which started the disorder, gave a bad name to all associations, the pugnaciously loyal Essex curate Thomas Twining announced with satisfaction. Essex's county's association for constitutional reform, he wrote on 14 July 1780, 'has been silent & inactive of late. I hope', he continued, 'that *associating* spirit will be quenched by the terrible effect of the Protestant association. It must open the eyes of many well-meaning people who had before joined in them.'[47] He was probably correct: the Protestant Association and the riots seem to have frightened property-owners and persuaded many moderates that the time was not ripe for introducing reforms in the representative system.

But if we are to conclude that the riots were more important than belated British military success in explaining the falling away of enthusiasm for reform, we should recognize that the riots themselves were linked to the war. Their immediate cause was the passing of a Catholic relief bill for England in 1778, and proposals for a corresponding measure for Scotland in 1779. These initiatives can be seen as part of a gradual process of softening of anti-Catholic feeling in elite circles – a process that perhaps started with the Quebec Act of 1774 (despite the opposition's squeals of protest and dire predictions), that was continued with the concessions to Irish Catholics in the 1790s, and culminated in Catholic emancipation in 1829. But the reforms of 1778–9 can also be seen as a direct response to the American war. North's ministry was inclined to support relief partly because conspicuous Catholic loyalty during the conflict was thought to merit recognition and reward, and partly because it was thought that relief would encourage further Catholic enlistment in the armed forces. And if the war influenced the parliamentary success of the reforms, it also conditioned the popular rejection of them. Struggles against the House of Bourbon had always been portrayed as wars against Catholic superstition, and once France and Spain entered the fray the virulent anti-Catholicism latent in many eighteenth-century Britons was perhaps bound to come to the surface.

Government and opposition had been united over Catholic relief: the bill of 1778 was introduced by Sir George Savile, an associate of

Rockingham. But co-operation was distinctly limited in this period; for the most part the war widened the gap between the opposition parties and Lord North's adminstration. Not that government and opposition formed two monolithic blocs. North's ministry was not fully united. North himself was less enthusiastic about the war than some of the other ministers; Barrington, the secretary at war, favoured naval rather than land operations against the colonists; and Lord Dartmouth, the American secretary at the outbreak of hostilities, was said to consider himself 'too old a *Whig* to approve' of the government's measures.[48] These differences were not as serious as they might now seem, however; for Barrington was so loyal that he put his personal preferences aside, and at the end of 1775 the ineffectual Dartmouth was replaced by Lord George Germain. Germain was much more committed to the vigorous prosecution of the war, and with North still lukewarm, Germain and Sandwich, the first lord of the admiralty, became the government's chief war ministers.

The opposition divisions were perhaps more serious. Rockingham and Chatham disagreed on American affairs – Chatham, indeed, seemed to be incapable of working closely with any other leading figure in the opposition. And even within his own party Rockingham had problems. The Rockinghamites were united in the belief that the war against the Americans was wrong – wrong because it would inevitably lead to a great expansion of the patronage powers of the crown which would endanger the British constitution, and wrong because it would destroy the prosperity of the empire and would therefore damage Britain commercially. But general agreement on these points masked considerable variations of opinion on tactics. While the marquis himself, in typically careful language, condemned the ministry's determination 'to pursue a Plan of Vindictiveness & Passionate Measures' without regard to the consequences,[49] some of his party were less cautious, and found themselves perilously close to hoping for the defeat of British arms. Charles James Fox was dismayed to hear of British victory at the battle of Long Island, while William Baker, another Rockinghamite MP, suggested that he was indifferent to the fate of the British forces in an unjust war: 'If the Utter Ruin of this Country is to be the Consequence ... I am the first to say, Let her Perish'.[50] The declaration of independence placed the Rockinghams in an awkward position; their aim, in Rockingham's own words, had 'always been to try to preserve a friendly union between the Colonies and the Mother Country'.[51] Only gradually did they embrace the policy of recognizing the reality of separation.

Saratoga was important in this respect, as in so many others. Burgoyne's surrender was a major setback for the ministry. North's support in the Commons declined as independent-minded back-benchers deserted him: his normal majority of more than 150 fell to less than 100. A reappraisal of policy was thought necessary with the entry of France into the war, and in 1778 North put forward a peace plan that involved Parliament's renunciation of its right to tax America. But Congress was in no mood to accept any terms short of a recognition of independence, so North's ministry pressed on with the war.

The new situation was not as immediately advantageous for the opposition as one might imagine. War against France and then Spain bolstered North's position and presented Rockingham's party with a fresh challenge – how could they criticize the running of the war without seeming to asssist Britain's traditional Bourbon enemies? Even so, North's position was unenviable. As British military prospects darkened, support for the government again slipped. Tentative approaches were made to members of the oppostion in an attempt to strengthen the administration and to divide the parliamentary forces arrayed against North. In 1779 Lord Gower and some of his followers, sensing a loss of nerve on North's part, decided to leave the ministry. During the winter of 1779–80 North did indeed come close to collapse. His resolve was maintained only by the forceful encouragement of the king and the determination of a small group of close colleagues.

The emergence of extra-parliamentary protest movements, in Ireland and more particularly in Britain itself, seemed to offer an opening for the opposition. In fact it damaged the new sense of unity between Shelburne, Chatham's political heir, and Rockingham, who since Chatham's death in 1778 had been co-operating in their attacks on the government. Shelburne was sympathetic to the reformers, and was willing to consider changes in the representative system, while Rockingham, who was certainly no radical, was disturbed by the more 'theoretical speculations' of the Yorkshire Association – especially their enthusiasm for shorter parliaments and 100 new county members.[52] They could agree on more limited measures of constitutional purification, however, and in Parliament the opposition decided to appeal to the discontented but conservative backbenchers by sticking to 'economical reform'. In February 1780 Edmund Burke introduced a bill to abolish sinecures and in April John Dunning, a member of Shelburne's group, carried a motion declaring that the 'influence of the crown has increased, is increasing and ought to be diminished'.

Dunning's victory greatly discomforted the government, but it turned out, at least for the time being, to be the high-water mark of opposition success. Burke's bill was slowly but surely emasculated, and opposition bills to prohibit government contractors from sitting in the Commons and revenue officers from voting in parliamentary elections were defeated.

North, still lacking confidence even though the opposition had clearly lost momentum, put out feelers for a coalition shortly after the Gordon riots. Rockingham's terms – the dismissal of Germain and Sandwich and recognition of American independence – were unacceptable, and North determined to soldier on alone. As news arrived of British success in South Carolina, North's colleagues urged him to capitalize on the upsurge in support for the government. In September 1780 he called a general election. The opposition, which seemed to have the ministry on the ropes only a few months before, made no progress; indeed, Burke lost his seat at Bristol as a result of his sympathy for the economic grievances of his Irish compatriots.

The opposition strongly condemned the declaration of war on the Dutch – the traditional Protestant ally – but not until news of Yorktown reached London on 25 November 1781 did North's position really begin to crumble. 'A general dispondency was the first effect of Lord Cornwallis's surrender', wrote Lord Loughborough, the lord chief justice. 'All the dissatisfied friends of govt. ... blame the system of the war ... & declare against its continuance'.[53] North's majority dropped to about 20. Germain was offered up as a sacrifice, and resigned in February 1782. Even so, on 27th of that month General Conway's motion to suspend the war in America was carried by 234 votes to 215. North was now under almost irresistible pressure to go. Finally, on 18 March, he was told by a group of normally loyal back benchers that they were about to withdraw their support. Rather than be defeated in a vote of no-confidence, he resigned.

The king with great reluctance accepted a new ministry headed by Rockingham and including Shelburne's group. The coalition was committed to negotiating peace with America and to introducing 'economical reforms' along the lines proposed by the opposition in 1780. Some progress was made in both areas, but on Rockingham's death in June 1782 the ministry disintegrated. Shelburne, the king's choice as Rockingham's successor, was detested by Fox, who took Rockingham's old following into opposition. Shelburne proposed to remodel the militia to create a nation in arms, he made noises about reforming the electoral system that encouraged his out-of-doors supporters, and he negotiated the peace – only to find himself the focus of parliamentary resentment at national humiliation. In February 1783 his ministry was twice defeated in the Commons and Shelburne, like North before him, decided that he could stay no longer. The American war can therefore be said to have destroyed not just one prime minister but two: North who paid the price for its unsuccessful prosecution, and Shelburne for concluding a necessary but unpopular peace.

The parliamentary divisions over the American war were echoed in the country at large. Even in the armed forces there were some dissident voices. Admiral Keppel refused to serve against the colonists, and the Earl of Effingham resigned his army commission rather than fight in America. Outside the army and navy the number of opponents of the war was much greater. In the early months of the conflict the pages of the officially published *London Gazette* were full of loyal addresses, seemingly vying with each other to express the bitterest repugnance of the Americans' conduct and the most obsequious praise of the king and his government; but at the same time there were counter petitions regretting the war and warning of its calamitous effects on British trade. Scotland, as we have seen, was conspicuously loyal – according to a report from the lowlands no more than 'one in a hundred of us ... has not espoused the side of Gov^{t}.'[54] - but, in many areas of England, army officers on the recruiting service met with lack of co-operation from local elites and resistance from local people. Lord Barrington went so far as to suggest that there was a danger of something like a civil war in Britain. He was particularly worried about the Wilkites in London, and opposed sending detachments of the Guards to America while there were so many 'desperate and ill-affected people' in the capital. An uprising in London, he warned, would soon spread to those other parts of the country where there was 'a very levelling spirit'.[55]

'Levelling' principles were no doubt associated in Barrington's mind with religious dissent, and dissenters of every complexion certainly tended to be pro-American. Heterodox Unitarians, however, were particularly outspoken critics of the war. On the eve of the conflict, Richard Price told Chatham that Britain was in great danger if Lord North crushed the colonists. As he wrote to another correspondent at the same time, 'By the government our ministers *endeavour* to establish in *New*-England, and that which they *have* established in ... Canada, we see what sort of Government they wish for in this country'.[56] In Price's imagination, it seemed that the Coercive Acts and the Quebec Act were merely the prelude to the adoption of a similarly authoritarian system of government in Britain. In 1776 Price published his *Observations on the Nature of civil Liberty, the Principles of Government, and the Justice and Policy of the War with America*, a highly controversial and frequently reprinted tract that anticipated many of the democratic arguments that were to circulate in the 1790s. The less well known Sylas Neville, an Edinburgh medical student of Unitarian inclination, was just as committed to the Americans and just as fearful of the consequences of their defeat. He fervently hoped that early British military progress – 'the triumph of the friends of despotism' was how he saw it – would be short lived. For Neville, as for

Price, if the government succeeded in 'making slaves' of the Americans, the British people themselves would be the next victims.[57] Like the rebellious colonists, dissenters of Neville's stamp pointed to a revival of Stuart-style tyranny: Samuel Kenrick was alarmed by the government's use of old Tory apologists – Dr Johnson, 'a most invincibly bigotted jacobite' and John Shebbeare 'who was pilloried above 30 years ago, for his jacobitical writings' and who had 'bespattered with the most virulent abuse' King William and the Glorious Revolution of 1688–9.[58]

Johnson and Shebbeare were both high Anglicans, and the enthusiasm of the dissenters for the American cause was probably reinforced by the stance of the established church. The government's requirement that Anglican clergymen preach loyal sermons on general fast days highlighted the intimate relationship between church and state, and so increased dissenting hostility to religious establishments. The content of these fast day sermons was not always confrontational. When churchmen attributed the wartime crisis to national degeneracy, and urged a general reformation of manners and morals, they were expressing views that few dissenters would have found offensive. Private vices such as 'Effeminacy and Frivolism', 'Gaming, Swearing, and Blasphemy' came in for particular condemnation,[59] and the sabbatarian act of 1781 that aimed to tighten Sunday observance, introduced into the House of Lords by Beilby Porteus, the bishop of Chester, can be seen as a tangible manifestation of this contemporary concern. But, much more controversially, Anglican sermons attacked the 'unnatural rebellion' in America,[60] and criticized so-called 'national sins' in Britain itself – namely 'want of Reverence for Governors', and 'a mutinous and licentious spirit',[61] which was blamed on the parliamentary opposition and, of course, on the dissenters.

The wickedness of the dissenters was indeed a favourite topic in the pronouncements of many leading Anglicans. 'Price', announced the future dean of Christ Church, Oxford, 'writes against all principles of Government.'[62] William Markham, the Archbishop of York, whose brother served as an army officer in America, was more outspoken. At various times he suggested that British victory should be followed up by the establishment of an episcopacy in America, and that there was a conspiracy against the divinely-ordained order entered into by dissenters on both sides of the Atlantic. Staunchly loyal provincial newspapers enthusiastically took up this theme. *Jopson's Coventry Mercury* likened the activities of the rebels in America to those of the dissenting mayor of the city. They were linked, the paper maintained, by a common commitment to Cromwellian fanaticism.

These references to the political and religious struggles of the seventeenth-century – like opposition and dissenting fears of a revival of Stuart principles – serve as a useful reminder of the extent to which those struggles shaped late-eighteenth-century political discourse. But if national disunity at the time of the American war was partly caused by a rekindling of old animosities, we should recognize that the divisions of this period anticipated the ideological polarity usually associated with the French Revolutionary War. To the supporters of Lord North's government the Americans were throwing off all restraints and, in the name of liberty, destroying the balanced constitution and establishing an unchecked democracy. To opponents of the government the Americans were legitimately defending the rights of Englishmen against a ministry with dangerously authoritarian tendencies. In the 1790s the pro- and antiwar factions divided along similar lines. Britons who opposed the conflict with revolutionary France argued that the French were merely establishing a system that embodied the true principles of the British constitution; Richard Price re-appeared in 1789 to claim that the French were only doing what the English had done in their own Revolution 100 years before. On the other side emphasis was placed on the dangerous way in which the French, in the name of liberty, were creating a regime that ignored established laws and oppressed its people. In denunciations of French republican despotism we can hear clear echoes of earlier protests against 'the tyranny of Congress'. To underline the continuities of argument, the Rev. Jonathan Boucher, an American loyalist who settled in England, decided in 1797 to publish a series of sermons he had delivered between 1763 and 1775. His message, in essence, was that the French Revolution was the 'giant offspring' of the American Revolution.[63] Edmund Burke, now in the conservative camp, was wrong, Boucher argued, to make a distinction between the disorders in France and those that had taken place in America. The French revolutionaries were repeating, if on a grander scale, the sins of their American predecessors.

Chapter 8 Notes

1. *Journals of Gilbert White*, ed. Walter Johnson (London, 1982 edn.), pp.156, 158, 164, 171, 177, 182, 187, 190, 198–9.
2. Round MSS, D/DRg 4/30, Essex RO
3. *An American Quaker in the British Isles: The Travel Journals of Jabez Maud Fisher*, ed. Kenneth Morgan (Oxford, 1992), p.229
4. Entry of 27 May 1782, Diary of Jonathan Adams, 547/1, Wiltshire RO.
5. Foreign Letter-book, 1776–8, Houston Papers, MS 8793, p.79, National Library of Scotland.

6. 'Correspondence of the Reverend Joseph Greene', ed. Levi Fox, Dugdale Society, *Publications*, xxiii (1965), 125.
7. Entry of 6 May 1778, Diary of Richard Neville, Neville and Aldworth Papers, D/EN F55/1, Berkshire RO.
8. Jolliffe to Jenkinson, 2 April 1779, WO 1/1004, p.395.
9. Lord Althorp to his mother, Lady Spencer, 21 Sept. 1779, Althorp Papers, F8, British Library.
10. Diary and Account-book of Matthew Flinders, 1775–84, fos. 29–30, Lincolnshire Archives Office.
11. Quoted in D.M. Joslin, 'London Bankers in Wartime 1739–84', *Studies in the Industrial Revolution presented to T.S. Ashton*, ed. L.S. Presnell (London, 1960), p.175.
12. *A Scottish Firm in Virginia 1767–1777: W. Cuningham and Co.*, ed. T.M. Devine (Edinburgh: for the Scottish History Society, 1984), p.212.
13. Quoted in T.M. Devine, 'Glasgow Merchants and the Collapse of the Tobacco trade 1775–1783', *Scottish Historical Review*, lii (1973), 74.
14. Rhinelander to Rawlinson & Chorley, 12 Nov. 1777, and to Smith Son & Russell, 29 Jan. 1778, Frederick and Philip Rhinelander Letter-book, 1774–83, New-York Historical Society.
15. Oliver's Journal, vol. i, Egerton MS 2,672, p. 84, and vol. ii, Egerton MS 2,673, p.565, British Library.
16. *The Diary of Sylas Neville, 1767–1788*, ed. Basil Cozens-Hardy (Oxford, 1950), p.280.
17. *The Selected Letters of Josiah Wedgwood*, ed. Ann Finer and George Savage (London, 1965), p.241.
18. Lt.-Col. James Bruce to Lord Ailesbury, 6 July 1778, Savernake MSS, 9, Wiltshire RO.
19. *The Correspondence of Jeremy Bentham*, vol. ii, ed. T.L.S. Sprigge (London, 1968), p.154.
20. Letter-book, 1778–83, William Wilson Papers, MS 9670, fo. 12, National Library of Scotland.
21. Quarter Sessions Order Book, vol. v, 1773–90, fo. 87, Worcestershire RO.
22. Matthew Lewis to Knowles, 21 April 1778, WO 4/102, p.351.
23. Capt. William Wood to Sir Matthew White Ridely, 30 Dec. 1776, Ridley of Blagdon MSS, ZR1 30/4a, Northumberland RO.
24. Journal of the Hon. George Napier, Napier Papers, vol. xvii, Add. MS 49,092, fo. 38.
25. Abingdon St Nicholas Parish Records, D/P2/17/1/1, Berkshire RO.
26. Liverpool Papers, Add. MS 38,343, fo. 290.
27. *Bedfordshire County Records: Notes and Extracts from the ... Quarter Sessions Rolls* (Bedford, n.d.), p.50.
28. Halkett Papers, MS 6410, fo. 26, National Library of Scotland.
29. Kenrick to James Wodrow, 29 Jan.–13 Feb. 1778, Wodrow-Kenrick Correspondence, MS 24.157 (59), Dr Williams's Library.
30. Forty-sixth Foot Muster Rolls, WO 12/5797, fo. 44.
31. Account of Green's Services, 7201–36, p.3, National Army Museum.
32. Gaol Book, City and County of York, 1775–85, Clerk of Assize Papers, ASSI 42/9, PRO.

33. Quoted in J.M. Beattie, *Crime and the Courts in England, 1660–1800* (Oxford, 1986), p.228.
34. Richard Laurence to George Wansey, 5 Aug. 1778, Wansey Papers, 314/4/1, Wiltshire RO.
35. Percy to Dr Thomas Percy, 7 Jan. 1776, Letters of Hugh, Earl Percy, MS G.31.39.7, Boston Public Library.
36. Lt. Charles Campbell to Alexander Campbell, 22 July 1778, Campbell Preston Papers, box 34, bundle 514, Ardchattan Priory.
37. *Ode on the Taking of Minorca* (London, 1782), p.7.
38. Althorp to Lady Spencer, 22 May 1778, Althorp Papers, F8.
39. John Moore to Arthur Annesley, 4 Oct. 1775, *Letters from an Ulster Land Agent, 1774–85*, ed. W.H. Crawford (Belfast, 1976), p.5.
40. Buckinghamshire to Sir Charles Thompson, 12 Oct. 1779, Hotham Papers, DD HO 4/20, Hull University Library.
41. HMC, *Twelfth Report*, Appendix, Pt. x (London, 1891), p.359.
42. *Pembroke Papers (1780–1794)*, ed. Lord Herbert (London, 1950), pp. 203–4.
43. Miscellaneous Papers, Add. MS 46,473, fo. 87.
44. Joanna Innes, 'Jonathan Clark, Social History and England's "Ancien Regime"', *Past & Present*, no. 115 (1987), 190 & n. 75.
45. Norcliffe to Christopher Wyvill, 7 Dec. 1779, Wyvill MSS, North Yorkshire RO.
46. Quoted in Linda Colley, 'Whose Nation? Class and National Consciousness in Britain 1750–1830', *Past & Present*, no. 113 (1986), 114.
47. *A Selection of Thomas Twining's Letters 1734–1804*, ed. Ralph S. Walker (2 vols., Lampeter, 1991), i. 184–5.
48. John Yorke to the Earl of Hardwicke, 29 July 1775, Hardwicke Papers, Add. MS 35,375, fo. 152.
49. Rockingham to Sir John Griffin Griffin, 22 Oct. 1775, Braybrooke MSS, D/DBy C9/17, Essex RO.
50. Baker to Edmund Burke, 22 Oct. 1777, Baker MSS, D/EBk C15/2, Hertfordshire RO.
51. *The Correspondence of Edmund Burke*, vol. iii, ed. George H. Guttridge (Cambridge, 1961), p.295.
52. Rockingham to Richard Croftes, 18 May 1780, Rockingham MSS, R1–1062, Sheffield Central Library.
53. Loughborough to William Eden, 13 Dec. 1781, Auckland Papers, Add. MS 34,418, fo. 213.
54. James Wodrow to Samuel Kenrick, 13 Aug. 1775, Wodrow-Kenrick Correspondence, MS 24.157 (52).
55. Shute Barrington, *The Political Life of William Wildman, Viscount Barrington* (London, 1814), pp.163–4.
56. *The Correspondence of Richard Price*, ed D.O. Thomas and Bernard Peach (2 vols. to date, Durham, N.C., and Cardiff, 1983–) , i. 189.
57. *The Diary of Sylas Neville*, pp. 245–6.
58. Kenrick to Wodrow, 12 May 1780, Wodrow-Kenrick Correspondence, MS. 24.157 (69).
59. James Scott, rector of Simonburn, *A Sermon Preached at York on the 21st of February, 1781* (York, [1781]), p.21.

60. John Warren, Bishop of St. Davids, *A Sermon Preached before the Lords Spiritual and Temporal, in the Abbey Church, Westminster, on Friday, February 4, 1780* (London, 1780), p.16.
61. Scott, p.18; Andrew Burnaby, vicar of East Greenwich, *A Sermon Preached before the Honourable House of Commons ... February 21, 1781* (London, 1781), p.14.
62. Rev. Dr. Lewis Bagot to Sir William Bagot, 14 March 1776, Bagot Papers, D 3259/13/18/4, Staffordshire RO.
63. *A View of the Causes and Consequences of the American Revolution; in Thirteen Discourses* (London, 1797), p.vii.

9

The Diplomacy of the War

Diplomacy and war are two different means by which governments seek to achieve their strategic objectives; but the two are intimately connected, and diplomacy does not stop once war begins. Official or unofficial negotiations often continue with open enemies; and when the belligerents refuse to talk to each other, they usually feel the need to talk with those who are not directly involved in the contest. Allies are sought and potential enemies have to be bought off. The story of any war, in short, is incomprehensible without a knowledge of its diplomacy.

The approach adopted here is necessarily selective. A detailed chronological narrative of every diplomatic twist and turn would fill a hefty book in itself. But we can cover at least a reasonable proportion of the vast expanse of ground by looking at the following themes: the failure of the various attempts to end the Anglo-American conflict between 1775 and 1778; Britain's search for allies; the making of the Franco-American alliance; Spain's entry into the war; the role of the Dutch; and, finally, the peace negotiations of 1782–3.

*

We have seen that attempts were made on both sides of the Atlantic in late 1774 and early 1775 to avoid military confrontation in the colonies. After the outbreak of hostilities at Lexington and Concord there were several further efforts to reconcile Britain and America, though none came very near to success. Perhaps the best chance was in the months between the start of the war in April 1775 and the Declaration of Independence in July 1776. During this period there were still a good many Americans who wished for a restoration of imperial harmony. Prominent amongst them was John Dickinson, who was the driving force behind Congress's so-called Olive Branch Petition to the king of 8 July 1775. This, in essence, was a general statement of loyalty and hope for better times; it offered no specific concessions and could not have formed the basis for negotiations. But, at the same time, Congress drew up an address to the British people, which contained a passage that might have provided grounds for a compromise settlement. The document suggested that if parliament dropped its claim to the right to tax America then the colonies would accept the Navigation Acts. Alternatively, in return for unrestricted trade, they would contribute to the costs of imperial defence. A similar hint was made when Congress formally rejected Lord North's conciliation proposal on 31 July 1775. Whether this was a serious offer, or merely a rhetorical device designed to demonstrate the reasonableness of the Americans, we shall never know. The British government was in no mood to consider it anyway. Divided sovereignty was as unacceptable to most British politicians in 1775 as it had been ten years before: 'if once they shall find themselves able to cut off one half of their subordination & dependence', an MP wrote, 'the other will soon follow after it let them promise what they will'.[1]

The next peace initiative was British, but not one sponsored or encouraged by the government. Admiral Lord Howe and his brother General Howe were appointed peace commissioners in May 1776. They were empowered to accept the submission of the colonists and to restore royal government. Then, and only then, were they permitted to negotiate with the pacified colonies (or any one or more of them) on the basis of Lord North's conciliation proposals of February 1775. But it seems that the Howes' sympathy for the colonists led them to go farther than the government intended. Lord Howe, in particular, chose to interpret his instructions as broadly as possible. Upon his arrival at New York he contacted his old acquaintance Benjamin Franklin. Howe received short shrift: 'To propose now to the Colonies a Submission to the Crown of Great Britain, would be fruitless', Franklin wrote. 'The Time is past.'[2] Howe was undeterred. As his secretary Henry Strachey observed, the admiral and his brother 'have nothing so much at heart as being the Instruments of reconciling the two Countries upon solid, advantageous, honorable Terms'.[3] Although Lord Howe's instructions did not permit him to negotiate with Congress (which the British government regarded as an illegal body), shortly after the battle of Long Island Howe offered to meet individual congressmen who might speak for the colonies. Howe let it be known unofficially that American grievances would be redressed and that, in due course, Congress would be recognized. After much discussion in Congress, Franklin, John Adams and Edward Rutledge were sent to confer with Admiral Howe. Adams, who was against any concessions and was hostile to the very idea of a meeting, correctly predicted that 'the whole will terminate in nothing'.[4] The discussions were, by all accounts, amicable enough; but even after a good meal Howe failed to persuade the American trio to renounce 'the System of

Independency'.[5] As Rutledge reported to Washington: 'our Conference with Lord Howe has been attended with no immediate Advantages .... Our Reliance continues therefore to be (under God) on your Wisdom & Fortitude & that of your Forces.'[6]

The Howes seemed to be more reluctant to turn their attention solely to military matters. Once Washington had been forced to abandon New York City, the brothers issued a proclamation to the American people on 19 September 1776 inviting discussion of reconciliation and stating that the king was prepared to allow a considerable measure of colonial self-government. Ira D. Gruber has argued that their attempt to blend negotiations and fighting was a serious error: 'strategy became hopelessly enmeshed with politics', to the detriment of the effective conduct of military operations.[7] The Howes, and Admiral Howe in particular, 'tried to use only enough force to create the illusion of British invincibility .... they sought the right combination of force and persuasion'.[8] Trenton and Princeton shattered the image of irresistible British power that the brothers had been carefully building, and so ruined their policy. But by this time, Gruber tells us, the Howes had succeeded in undermining the government's own policy of settling the issue by purely military means. By refusing to go for the jugular in the 1776 campaign, the Howes had thrown away the best chance of decisive victory.

Reconstructing the intentions of the Howes, as Gruber readily admits, is no easy matter. Very little of their private correspondence has survived. But Gruber has examined a vast amount of relevant material and he presents his case convincingly. Contemporary testimony seems to bear out his claim that the Howes held back from delivering a telling blow. We have seen, for instance, how Captain William Bamford of the Fortieth Foot complained of the opportunities that the British had 'let slip' at Brooklyn and White Plains; and how the Hessian Captain Johann Ewald hinted darkly at the motives behind General Howe's reluctance to press the Americans after the battle of Brandywine.[9] Likewise, Sir George Collier, one of Lord Howe's naval officers, implied that the failure to cut off Washington's retreat from Brooklyn was attributable to the admiral: 'had our ships attackd the batteries, which we have been in constant expectation of being orderd to do, not a man could have escapd from Long Island'.[10]

A few caveats and qualification might be in order, however. Bamford, we should recall, seemed to impute the restraint of General Howe to the influence of senior officers in the ancillary services – officers who wanted to prolong the war to increase their profits. Right or wrong, his accusation reminds us that there might have been a variety of considerations, besides the political ones that Gruber identifies, that inhibited the vigorous pursuit of the Continental army. A supply and communication line stretching over 3,000 miles of ocean was surely one. The knowledge that reinforcements would take many months to arrive can hardly have encouraged boldness either. Nor should the carnage of Bunker Hill be forgotten. It seems extemely unlikely that the appalling casualties incurred by the British on 17 June 1775 had slipped out of General Howe's memory by the time he started the 1776 campaign at New York: after all, Bunker Hill was then the last battle that he had fought. At the very least it must have given him a keen appreciation of the diffculties involved in shifting the Americans from defensive works. The more recent British disaster in South Carolina might have been influential, too. Strachey, who was close to the Howes, was fearful before the battle of Long Island that the British might 'suffer such another Disgrace as at Charles Town'.[11] We should also note that a naval expert has defended Admiral Howe's conduct after the battle of Long Island as a 'logically justifiable approach'. Lord Howe had no reason to believe that the defeated Washington was about to retreat across the narrow waters between Brooklyn and Manhatten: on the contrary, Washington had just called for reinforcements and he seemed determined to stay on Long Island. For Lord Howe to have ordered his ships to patrol the waters on the off-chance that Washington might choose to withdraw would have involved considerable risks, not least from the fire of American shore batteries. A recent attack on some of his vessels by American fireships might have made the admiral still more sensible of the danger of 'exposing big frigates merely for patrol purposes'.[12] None of this, of course, fatally undermines Gruber's thesis. His version of events remains attractive. But it may only partly explain why the Howes failed to accomplish what the British government and the American loyalists expected of them.

As the Howes were tilling the stoney ground of negotiations, a less well known ploughman was labouring with the same lack of success. Major Apollos Morris of the Twenty-seventh Foot was one of the small number of British officers opposed to coercion of the colonies. In September 1775 the commander-in-chief in Ireland strongly recommended that the major be given permission to sell his commission and retire: 'He ... found fault with this most just & necessary War, ... And when I would have interrupted him, He thundered out an hundred Greek Lines out of Homer – He then talk'd to me out of Plutarchs Lives ... he will be better out of the King's service than in it'.[13] At the end of the year, having left the army, Morris wrote to Germain, arguing 'tis certainly in our Interest to put a conclusion to the War'.[14] But Morris was not content with tendering advice to the new American secretary. In 1776 he sailed for America. His motives are not entirely clear. Major Christopher French of the Twenty-second Foot, a prisoner with the revolutionaries, claimed that Morris had 'acknowledg'd he had come to America upon promise of a Reg^{t} & being appointed Adjutant General to the Rebel Army'.[15] Washington certainly considered him for the post, and Morris himself said that he had originally been prepared to take up arms to defend America's 'Peace, Liberty & safety', which he believed could best be secured by a return 'to the situation of 1763'.[16] But once he discovered that the Americans were now contending for something rather different, he was less willing to fight his former colleagues. Instead of serving in a military capacity, he tried to advance the peace process. In January 1777 he wrote to Lord Howe, proposing a cessation of hostilities while the commissioners sought full authority to negotiate, and to Washington, urging him to 'add, to the honours you have already gained', by promoting an accommodation. Morris was still convinced that the Americans needed only to agree to contribute to the British exchequer: 'some offer not only of a mutually advantageous commerce but of taking a part of the national debt *when able'* would surely solve the difficulty.[17] Like Lord Howe before him, Morris could not make the necessary mental adjustments to come to terms with American Independence. The issues at stake had changed. Arrested by order of Congress in June 1777, he was expelled from the country.

There was no further British initiative – official or unofficial – until Burgoynes's defeat at Saratoga and the prospect of the entry of France into the war obliged North's government to reconsider its American policy. An emissary was sent to Paris to sound out the American agents who were already negotiating with the French court. Nothing much came of this tentative approach; but, with the aim of ending the war before France could effectively intervene, North rushed two bills through Parliament that were designed to bring the Anglo-American conflict to an end. The first of these bills offered the Americans almost all they had demanded in 1775: repeal of the Coercive Acts and freedom from parliamentary taxation. The second created a new commission to negotiate on the basis of Parliament's concessions. There were some grounds for optimism. Although the Earl of Carlisle, the nominal head of the commission, was no intellectual heavyweight, William Eden, its driving force, was the author of a treatise on penal law and a politician and administrator of high calibre. One of the secretaries to the commission, furthermore, was Adam Ferguson, the Edinburgh philosopher. But hopefulness was in short supply across the Atlantic. From the British base at Philadelphia Captain William Duff greeted news of the conciliation bill with dismay: how could it be expected, he asked, that those American leaders who had acquired power at local or national level 'will be prevailed upon to relinquish it, and return to their original obscurity'?[18]

Major Alexander Dundas, another British officer, predicted that Congress would 'refuse to treat with the Commissioners'.[19]

Carlisle and his colleagues were soon similarly despondent. They received the first blow to their morale when they arrived at Philadelphia. Orders to evacuate the city had been sent from London without informing the commissioners, and they felt that their negotiating position had been severely weakened. 'As long as we had the *army* to back us ... we had hopes of success', Carlisle told his wife on 21 June, 'but this turning our backs on Mr. Washington will certainly make them reject offers that perhaps fear of what that army could have done would have made them listen to.'[20] Attributing their difficulties to the evacuation of Philadelphia no doubt consoled the commissioners for their failure. But the real problem lay with the terms that they were empowered to offer. The concessions of 1778 would probably have satisfied the Americans in 1775; but North's government was not prepared to concede immediate withdrawal of all British troops from the United States or to recognize American Independence, so the Americans refused to talk. In late July Carlisle learned – via a report in a Philadelphia newspaper – that Congress had decided to ignore the commissioners. Such a contemptuous rebuff was perhaps inevitable. Too much had happened since the outbreak of war for Congress to contemplate any turning back of the clock 'to the Situation of 1763'. As George Mason explained in October 1778, more than three years of fighting had left a legacy of deep bitterness:[21]

> Can they raise our Citys out of their Ashes? Can they replace, in Ease & Affluence, the thousands of Familys whom they have ruined? ... Our Country has been made a scene of Desolation & Blood ... We can never again trust a People who have thus used us .... The Die is cast – the Rubicon is passed – and a Reconciliation with Great Britain, upon the Terms of returning to her Government is impossible.

*

If the British tried to negotiate with the Americans, they also attempted to secure allies, especially once the French became open enemies. They were as unsuccessful in this aspect of their diplomatic activity as they were in their approaches to the estranged colonists. We have already seen (in Chapter 3) that advances to Russia and the United Provinces in 1775 met with a cold response. In February 1778, shortly after North's government learned of the signing of the Franco-American treaties, another approach was made to Catherine the Great. But the British were unwilling to pledge military assistance in the case of a Russo-Turkish war. As such military assistance was the only possible advantage Catherine could see for Russia from any arrangement with Britain, she rejected the British offer of an alliance. Prussia appeared a more promising prospect in April 1778. A German crisis was brewing over Austria's territorial ambitions – Emperor Joseph II coveted Bavaria – and Frederick II was looking for friends in the event of a confrontation. George III, in his capacity as elector of Hanover, was well disposed towards a Prussian alliance in return for a guarantee for the electorate. Prussia, however, lost interest once it became clear that France and Russia had no intention of intervening on Austria's behalf. With Prussia and Russia disinclined, Austria seemed the only remaining option. In December 1780 the British hoped to court Joseph by promising support for the opening of the river Scheldt to international navigation, a development likely to benefit the Austrian Netherlands. When this failed to woo Joseph, further efforts were made to win Catherine's friendship. In a desperate last throw in March 1781 she was even offered the Mediterranean base of Minorca, but to no avail. Catherine, the British ambassador reported, was convinced 'that we wanted ... to *draw her into the war*'.[22] Both the Austrians and the Russians were happy to act as mediators – and, indeed, performed this part in the final negotiations – but neither was prepared to befriend Britain in her hour of need.

Historians have long criticized the conduct of British foreign policy before and during the American war.[23] Ministerial unwillingness to offer sufficiently tantalising inducements is seen as the cause of Britain's damaging isolation. France, as a result of this diplomatic failure by successive British ministers, was not faced with the distraction of a major European conflict, or even held back by the consideration that Britain had a friend or friends in Europe who might come to her assistance. Free from such worries, France could concentrate her resources on a war of revenge.

This may be an overly harsh judgement. The enormous debt incurred in the Seven Years' War loomed large in the minds of every British government after 1763, and made them anxious to keep public spending under control. Financial considerations inevitably limited the scope for diplomatic manoeuvre; a more interventionist foreign policy would have cost a good deal of money. In any case, lack of allies was not an obvious impediment for Britain prior to 1775, or even 1778. Britain's naval superiority ensured that she was able to compel the French to climbdown in the crisis of 1764 over Turks Island in the Caribbean, and the Spanish were also forced to retreat over the Falkland Islands in 1770–1. Once the American war started, allies would no doubt have been useful, if only to encourage the French to glance nervously over their shoulders every now and then; but the government can hardly be held solely responsible for failing to win friends. Britain had little desirable to offer Austria, Prussia and Russia, all of which were increasingly preoccupied with central and eastern Europe. It was not, as the attorney general told the House of Commons in June 1779, in the power of the government 'to oblige foreign courts to enter into alliance with Great Britain'. Indeed, he continued with remarkable candour, given that 'the necessity was solely on our side ... it was not at all to be wondered at, that other courts were not much inclined to enter into alliance with us'.[24]

*

The 13 colonies recognized that they needed foreign assistance as soon as the war began. European money, supplies and military advisers were vital if the British army and navy were to be successfully resisted. France, Britain's great rival, was the obvious power to approach – even if her Catholicism and absolutist system of government made her less than a natural ally. The French, according to a British visitor in September 1775, 'are universally favourable to the Americans',[25] not least, one suspects, because they were happy to see the British in difficulties. More important than popular sympathy for the rebellious colonists was the willingness of Vergennes to oblige the Americans. As early as August 1775 he authorized the sending of an agent to confer with Congress at Philadelphia. Opinion within the French government was divided, but by the spring of 1776 Vergennes had prevailed over the advocates of caution, led by Turgot, who wished to concentrate on domestic reform and feared that a fresh conflict with Britain would be ruinously expensive. In May Louis XVI agreed to financial aid for the Americans. To keep the involvement of the French government secret, a company (Roderigue Hortalez et cie) was established to use the money to purchase arms and ammunition from the government's own arsenals. The munitions were to be sold to Congress on the credit of anticipated tobacco shipments to Europe. Clandestine as these arrangements were, the likely outcome was clear enough for Turgot to resign shortly after Louis's decision to support the insurgents. Released from Turgot's restraining influence, the French government began to plan for war.

By September 1776 Congress had drawn up a model treaty of amity and commerce to offer to France and Spain. Franklin and Arthur Lee were appointed commissioners to join Silas Deane (who was already in France to purchase war supplies) and present the draft treaty to Vergennnes. In January 1777 the American commissioners reported that 'The Hearts of the French are universally for us, and the Cry is for immediate War with Britain'. They added, however, that 'the Court has its reasons for postponing it a little longer'.[26] The French were still a long way behind the

British in numbers of ships of the line, and Spain had to be persuaded to join the conflict if the Royal Navy was to be fully stretched. In the meantime, it was in the interest of the French to avoid open hostilities. The use that American privateers made of French ports naturally led to a deterioration in relations with the London government, and to appease the British Vergennes even went so far as to imprison the captain and crew of one of the American vessels at Dunkirk.

News of Burgoyne's capitulation did not, then, change the direction of French policy. Nor, as sometimes is suggested, did Washington's doughty performance at Germantown exert a great influence. France was already moving towards war with Britain. But events in America did make some difference. Saratoga in particular gave Vergennes the opportunity to overcome the king's remaining doubts and reservations. The British now looked more vulnerable, and Vergennes was able to frighten Louis with rumours that the British and Americans might come to terms and join in an attack on the French West Indies. A few days after news of Burgoyne's defeat reached France, representatives of the French government met the American commissioners. The commissioners reported that 'after long and full Consideration of our Affairs and Propositions ... his Majesty was determined to acknowledge our Independence and make a treaty with us of Amity and Commerce'. The king, moreover, 'was fix'd in his determination, not only to acknowledge but to support our Independence, by every means in his Power'. Louis recognized that this would probably mean war, yet he was now willing to fight because 'besides his real Goodwill to us and our cause, it was manifestly the Interest of France that the Power of England should be diminished by our Separation from it'.[27]

On 6 February 1778 the treaty of amity and commerce and a Franco-American alliance were signed. 'Our States have now a Solid Support for their Liberty and Independence', Franklin and his colleagues wrote proudly.[28] In March the British brought their ambassador home, and the French responded by withdrawing their minister from London. The French would have preferred to delay the outbreak of fighting until they could bring in the Spanish. But the Spanish had no particular interest in American independence, and they were determined to extract as much advantage as possible from their strong position. The French government therefore took the risky course of proceeding without the Spanish. They placed their faith in an immediate and decisive victory. Admiral d'Estaing and the Toulon fleet were despatched to America in April to trap Admiral Howe and compel the surrender of the British army at New York. To prevent the early departure of British naval reinforcements to North America, French troops were massed along the Channel coast to suggest that an invasion was about to be launched. This last part of the plan worked; as we have seen, the British admiralty delayed sending ships to help Howe until there was no doubt that d'Estaing was sailing across the Atlantic rather than joining the Brest fleet for a descent on the south coast of England. But d'Estaing was unable to snare Howe. Having failed to deliver the blow that might have ended the war, the French desperately needed Spanish help for the campaign of 1779.

*

The Spanish, like the French, were keen to recover territory lost to the British in 1763. The return of the Floridas and control of navigation on the Mississippi were seen as vital to the well-being of the Spanish possessions in the Caribbean and North and Central America. But Spanish grievances went back a lot farther than the Peace of Paris. British settlers at Belize in the Bay of Honduras and on the Mosquito Coast of Nicaragua had been an irritant for much of the century; and Spain still had hopes of reclaiming Jamaica, seized by the English in 1655. In Europe, Gibraltar had been in British hands since 1704 and Minorca since 1708. The return of these Mediterranean bases was of the utmost importance to Madrid. But if Britain, as the occupying power, was the obvious obstacle to the achievement of these aims, it did not necessarily follow that war with Britain was the only way to fulfil Spain's ambitions. Spain was prepared to get what she wanted by whatever means were required: either by negotiations with London or by joining the French in an alliance directed against the British.

It was clearly in the interests of Britain to keep Spain neutral, just as it was vital for the French to bring her into the war. To take full advantage of this strong position, in 1778 the Spanish government offered its services as mediator. The expectation at Madrid was that the British and French would then bid against each other for Spain's favour. The British had already shown a willingness to appease the Spanish. Between 1774 and 1777 the London government tried to promote a settlement of the dispute between Spain and Portugal over the borders of Brazil. As long-standing allies of the Portuguese the British might have been dragged into a conflict with Spain (and therefore with France) just when they were trying to concentrate their efforts on putting down the American rebellion. Rather than extend the war at this crucial juncture, the British put pressure on the Portuguese to come to terms. But in 1778–9 the British were not themselves prepared to make the necessary concessions. An offer of Gibraltar and Minorca would almost certainly have sufficed to secure Spanish neutrality. Yet no such offer was seriously contemplated.

In part this was because the British government, influenced no doubt by the army and navy, was convinced of the inferiority of the Spanish armed forces (a Spanish war, wrote a marines officer in June 1779, 'is always agreeable to the Jack Tars of Britain; who always lay their account with giving the Proud Spaniard a sound Drubbing; as witness last war';[29] 'the Spaniards in my opinion [are] no great adepts in the management of fleets',[30] was the equally complacent comment of Admiral Sir Thomas Pye). But even supposing the British government had thought more highly of the Spanish, it seems unlikely that Parliament and public opinion would have tolerated the surrender of the Mediterranean outposts.

The French were much more willing to accommodate Spanish wishes. So desperate were they to expedite Spanish entry into the struggle that they agreed to almost every Spanish demand – even the project for an immediate joint attack on England designed to bring the war to a speedy conclusion. On 12 April 1779 the Spanish and French signed a formal alliance at Carlos's summer palace. The Spanish agreed to assist the French to gain the Newfoundland fisheries, Senegal in West Africa, Dominica in the Caribbean, and a better position in India. Spain, however, drove a hard bargain for her support. The French were obliged to promise not to make peace until the Spanish had taken Gibraltar, and to assist with the recovery of Minorca and Spain's lost American territories. Vergennes was distressed by the harshness of the terms – he had told Louis in December 1778 that 'the views and pretentions of Spain are gigantic'[31] – but he recognized that France needed Spain more than Spain needed France.

Spain's involvement in the war did not stop her keeping open her lines of communication with the British. Carlos's ministers had banked on a short and successful conflict that would minimize the risk to their far-flung and vulnerable empire. The failure of the invasion project of 1779 inclined the Spanish government to respond favourably to British peace feelers in early 1780. But the negotiations entered into by the British envoy Richard Cumberland came to nought. The chance to remove Spain from the coalition ranged against Britain was allowed to pass. Gibraltar was again the stumbling block. With the Spanish siege already underway, the British were even more reluctant to succumb. Public opinion would surely have never forgiven the ministers for handing over a post that the British army and navy were so successfully defending.

The Spanish contribution to American independence is now widely recognized by historians.[32] The victorious campaigns of de Galvez along the Gulf coast and the vital part played by Spanish ships in tipping the naval balance against the British have been discussed in earlier chapters.

Spanish financial aid to the new republic has also been mentioned. Spanish munitions assisted the revolutionaries, too. Yet, for all this, Hispano-American relations were far from cordial. Before they joined the war, the Spanish, in their role as mediators, had proposed a compromise settlement that would have left the British in possession of New York. Perhaps the offer was not meant to be taken seriously, but it was symptomatic of the Spanish government's reluctance whole heartedly to support the Americans. Spain entered the war as an ally of France, not of the United States. True, the Spanish agreed to make peace only with French consent, and since the French were committed to fighting on until the British recognized American independence, the Spanish were indirectly committed to the same goal. But this did not mean that they approved of the Americans. Far from it. A great imperial power like Spain had good reason to be wary about encouraging colonial rebellion – especially in territories so close to her own. Conde de Floridablanca, the secretary of state, was well aware that a strong and confident United States could pose an even bigger threat to Spain's American possessions than Britain had done in the past. Indeed, the Spanish rightly feared the westward expansion of the new republic and American control of the Mississippi. When in 1780 John Jay arrived at the Spanish court as minister plenipotentiary of the United States, he met with a decidedly cool reception. His attempts to negotiate an alliance were embarrassingly unsuccessful. Floridablanca's evasion and haughtiness drove Jay to distraction: the American complained bitterly of 'the littleness' of the Spanish attitude towards him and his country.[33]

*

For the British to add the Spanish to their enemies was unfortunate but perhaps unavoidable. To add the Dutch seems less excusable. Relations between the British and the Dutch were first strained by the refusal of the United Provinces to accept that the American war obliged them to assist the British under the terms of the Anglo-Dutch alliance of 1678. The republic, as we have seen, turned down a British request for the use of the Scots Brigade. Worse still, from the British perspective, was the extensive trade carried on by the Dutch with the American rebels, especially via the Caribbean island of St. Eustatius. Complaints from Sir Joseph Yorke, the British ambassador, did little to stem the flow of goods. The sheltering of John Paul Jones at the Texel in October 1779 seemed merely to confirm that the Dutch, despite their professions to the contrary, were in reality determined to help the Americans. But the anger and bitterness was not just one-sided. The Dutch were aggrieved at the Royal Navy's seizure of their merchant vessels in apparent contravention of the treaty of 1674, which was supposed to guarantee the principle of 'free ships, free cargoes' – or, in other words, to allow either party to carry goods (other than carefully defined contraband) which belonged to, or were destined for, a country with which the other was at war.

The outbreak of the Anglo-French conflict exacerbated the problem. The British were convinced that the Dutch were propping up the French war effort by bringing them vast quantities of Baltic naval stores which were vital to France's ship-building programme. This commerce was thought to justify still more searches and seizures by the Royal Navy. The crisis came to a head when the British feared that the Dutch were about to gain the protection of Catherine the Great and the Russian fleet. Catherine had been deeply offended when two Russian merchant ships in the Mediterranean were seized by the Spanish besiegers of Gibraltar. Her first reaction was to punish Spain, but her foreign minister, Count Panin, persuaded her that the problem was the unsatisfactory state of international law rather than Spanish aggression. Catherine accordingly announced in February 1780 her intention to form a League of Armed Neutrality to secure, by force if necessary, the rights of neutral shippers. Central to the provisions of the proposed league was the claim that naval stores should not be included in the very limited number of items considered as contraband.

The British were understandably alarmed. If the Dutch were to follow the Scandinavian powers and join the league, it would be much more hazardous to intercept their merchant vessels. Conflict with Russia, and the other members of the league, was certainly not what Lord North's beleaguered government wanted. Another concern was that the Dutch, protected by the other signatories of the league, would be able to carry even more enemy stores, to the enormous advantage of the Bourbon powers. And Lord Stormont, one of the secretaries of state responsible for foreign affairs, believed that if the Dutch joined the league, they would be able to handle most of the overseas trade of France and Spain, enabling those powers to use nearly all their available mariners on board war vessels. In December 1780 the ministers in London therefore used evidence of a draft treaty of commerce between the Americans and Jan de Neufville, a prominent Amsterdam merchant, as a pretext for a declaration of war. The parliamentary opposition condemned the government's action, but it seems to have been generally welcomed. In part this was because a Dutch war was likely to be profitable: Portsmouth was said to be 'mad with joy at the prospect' and to be 'full of people from London, come down to buy ships for privateers'.[34] But support for the government's action went far beyond the ranks of those who had an eye to material gain;

there were many Britons who resented what they saw as the secret hostility of a former ally: there was much criticism of 'the perfidious Dutch'.[35]

The British government secured its immediate objective in as much as Catherine refused to come to the assistance of the Dutch now that they were belligerents rather than neutrals. The Dutch war, moreover, was generally successful for Britain: St. Eustatius was taken, as were Negapatam in India, Trincomalee in Ceylon and the *VOC*'s bases on the west coast of Sumatra. The French recapture of St. Eustatius and Trincomalee took the shine off these triumphs, but at least the French were obliged to commit troops and ships to the defence of the remaining Dutch possessions; a French garrison stayed at the Cape of Good Hope until the end of the war.

Yet the advantages for the British were largely illusory. If the French had to commit forces to help the Dutch, the British – who were already more stretched anyway – also had to divert resources that might have made a real difference elsewhere: we have seen how the outbreak of the Dutch war led to the abandonment of the expedition against the Spanish possessions in the Pacific; it also obliged the British to commit ships and men to the capture and retention of Trincomalee, a port from which the French navy might operate on both the Coromandel and Malabar coasts of India. The distraction of a Dutch war might still have been worthwhile if the main aim behind it had been achieved. It was not. The British failed to cut off the flow of war supplies to the French, the Spanish and Americans. The Russians and the Prussians filled some of the gap left by the Dutch carriers, and the Dutch themselves continued to trade under the protection of foreign flags. Worse still, the British had overstated the importance of seaborne imports to the French navy. Large quantities of naval stores had been stock-piled before July 1778 as part of France's lengthy war preparations. And once the fighting started, there were other ways in which essential material could enter the country. The Dutch themselves sent goods through the canal network of the Austrian Netherlands. Italy was also a source of supply.

Nor, as we shall see, was another British objective fulfilled. It was hoped in London that a successful war against the Dutch would strengthen the position of the pro-British *Stadhouder* and weaken his Francophile opponents.[36] Instead, the struggle undermined the *Stadhouder* and pushed the Dutch into the arms of the French.

*

The fall of Lord North in March 1782 brought into office an administration under the Marquis of Rockingham which was committed to ending the war with America. Within the ministry, however, the was no agreement on the way forward. Fox, the occupant of the new post of foreign secretary, wanted to grant American independence without delay in order to pursue the struggle with Britain's European enemies with more vigour. But Shelburne, the secretary of state for home and colonial affairs, claimed that negotiations with the Americans came within his jurisdiction, and he was anxious not to recognize independence until a comprehensive peace settlement had been secured.

It would have been difficult enough for the new government if these policy differences were confined to cabinet meetings. But they were advertised to the world when the two ministers each appointed a representative to negotiate in Paris. Franklin and Shelburne had been in contact even before North's departure, and as soon as he came into office Shelburne hurried his envoy, Richard Oswald, across the Channel. Fox, not to be outdone, responded by sending Thomas Grenville to talk with the Americans and the French. Franklin was more interested in Oswald – an old acquaintance – and Grenville made little headway with Vergennes, not least because Fox insisted on using the peace of 1763 as the basis for Anglo-French discussions. Although Fox was in a minority in the cabinet, and opposed by the king, the crisis was resolved only when Rockingham died in July 1782 and Shelburne became first minister. Fox resigned, leaving Shelburne as the chief British player in the negotiations.

Franklin encouraged Shelburne to believe that generosity to America would pay dividends for the British. If the Americans appeared to be on the verge of agreement with the British, Franklin reasoned, the French and Spanish would become more amenable. He even suggested that Canada be given to the United States as a token of British goodwill. Shelburne was not prepared to go that far, but he was willing to give the Americans more than they could realistically have expected. Besides hastening the general peace by compelling the Bourbons to accept terms, he had in mind future Anglo-American relations. Strong trading links with the former colonies would help to boost Britain's commerce and manufacturing; and, just as important, British friendship with the United States would help to ensure that the Franco-American alliance lapsed with the peace. There was even the possibility that the Americans might enter into some form of political relationship with Britain.

Negotiations stalled when the accommodating Franklin fell ill and John Jay took over as the American representative. Jay was deeply suspicious of British motives, distrusted the French, and caused difficulties over Oswald's credentials. With Franklin's return, however, progress was resumed. Agreement between the British and the Americans was reached on 30 November 1782. By the preliminary terms (they were to remain provisional until the conclusion of a general peace) the British conceded the independence of the United States and pledged to withdraw their troops from the territory of the republic. The United States was given the lands between the Great Lakes and the Ohio, and fishing rights off Newfoundland. In return, the Americans agreed to honour their debts and promised that Congress would recommend to the states a restoration of British and loyalist property.

By this stage, Anglo-French negotiations were reaching a critical point. The French had initially wanted to reverse as much of the peace of 1763 as possible. Vergennes said that its mere mention was enough to make him 'shudder'. It was necessary, he wrote in June 1782 to comte de Montmorin, the French ambassador at Madrid, 'to forget all that had gone before', to start afresh and construct a new settlement 'based on justice and mutual convenience'.[37] Talks between Shelburne and the captured Admiral de Grasse in August 1782 suggested to Vergennes that the British were likely to make at least some of the concessions that the French desired. Vergennes sent his under-secretary, Gérard de Rayneval, on a secret mission to verify de Grasse's report. At a meeting on 13 September Shelburne denied the accuracy of de Grasse's version of their discussions: 'When your Lordship reads it and recollects the account I gave you of what pass'd,' Shelburne wrote to his foreign secretary, Lord Grantham, 'you will not need to be told, how much French imagination has added.' Shelburne refused to confirm the concessions he was said to have offered; on India, for instance, he explained to Rayneval that 'things must remain there as they were established in 1763', for 'it was not to be expected that the King could cede two continents'.[38] The outcome of the interview was not entirely negative, however. Shelburne put it to Rayneval that a reconciled Britain and France might co-operate and become the arbiters of Europe.

Vergennes was under pressure from his colleagues not to give ground to the British – the marquis de Castries, the minister of marine, was particularly anxious to win extensive concessions in India – but the possibility of British help in countering the Russians in Eastern Europe was tempting. Ever since he had known of the Austro-Russian alliance of June 1781, Vergennes had been convinced that Catherine would soon move against the Ottoman empire, a long-standing friend of France. At this juncture – in late September 1782 – his fears must have seemed fully justified: reports had reached him from the French ambassador at St. Petersburg of preparations for war against the Turks (in the autumn Russian troops occupied the Crimea). But there were other pressing reasons for Vergennes to adopt a more flexible approach towards the British. Rodney's victory at the Saintes offered worrying evidence of a British naval revival (Stormont, now out of office, astutely observed that Rodney had restored his country 'to such a Situation as will enable it to retrieve most of the Losses it has sustain'd').[39] 'The English', as Vergennes told his ambassador at Madrid, 'have to some degree regenerated their navy while ours has been used up.'[40] French financial exhaustion was no doubt also a major consideration. As early as September 1780 Necker, the finance minister, had pointed to France's perilous financial situation as a cogent reason for pulling out of the war. Under pressure from Necker, comte de Maurepas, then first minister, had even made a tentative but unsuccessful approach to Lord North in the preceding July. While Necker was an Anglophile and ill-disposed to the war from the beginning, not even an Anglophobe foreign minister could ignore France's financial plight.

Vergennes, then, was now prepared to compromise to bring the war to an end. Spain, however, notwithstanding her own financial difficulties, was unwilling to make peace without Gibraltar. Even after the defeat of the great attack of September 1782, the Spanish still coveted the rock. The Spanish minister in Paris, conde de Aranda, even suggested that the French could have Spain's Caribbean colony of Santo Domingo if they could persuade the British to cede Gibraltar in return for an equivalent piece of French territory. For a while Shelburne was prepared to consider giving up Gibraltar in exchange for extensive Spanish concessions (Minorca, Florida and Puerto Rico, or two French West Indian islands instead of Puerto Rico). But in December, after the signing of the Anglo-American preliminaries made the British less inclined to give more ground, the Spanish finally dropped Gibraltar from their demands.

The Dutch presented a less formidable obstacle to peace, but an obstacle all the same. Their emissary, Gerard Brantsen, was instructed to make no concessions; the United Provinces, having fared disastrously in the war, unrealistically sought the return of all their territories and trading bases, confirmation of the rights of neutral shippers, and an indemnity for their losses at sea. Vergennes, negotiating on their behalf, secured the return of the valuable port of Trincomalee; but the British were determined that in return they should retain Negapatam. The Dutch held out for as long as they could, only consenting to an armistice when the French and Spanish reached preliminary agreement with the British on 20 January 1783, and refusing to sign a final treaty until 20 May 1784. In the end, however, they achieved nothing by this delay. The British kept Negapatam and won the right to navigate (and by implication illicitly to trade) in the Dutch East Indies.

Meanwhile, the final treaties between the other belligerents were signed in Paris on 3 September 1783. They confirmed all the terms agreed in the preliminaries. The Spanish gained Minorca and the Floridas, but not Gibraltar; the French had to settle for a restoration of their prewar trading stations in India, but regained St. Lucia and retained Tobago in the Caribbean, recovered Senegal in Africa, secured better fishing grounds off Newfoundland, and, symbolically, won back the right to fortify Dunkirk. The British had lost the 13 colonies, widely regarded as the cornerstone of their empire and one of the chief sources of their national prosperity. But they had saved most of their West Indian possessions, and in India, despite the precariousness of their military position, they had successfully resisted French pressure for territorial concessions and had even strengthened their hold at the expense of the Dutch.

Chapter 9 Notes

1. John Yorke to the Earl of Hardwicke, 12 Aug. 1775, Hardwicke Papers, Add. MS 35, 375, fo. 155.
2. *Franklin Papers*, xxii. 575.
3. Strachey to ..., [26 Aug. 1776,] (draft) Strachey MSS, DD/SH, C.1165, box 34, Somerset RO.
4. *Papers of John Adams*, ed. Robert J. Taylor (8 vols. to date, Cambridge, Mass., and London, 1977–), v. 20.
5. *Franklin Papers*, xxii. 604–5.
6. *Letters of Delegates to Congress, 1774–1789*, ed. Paul H. Smith (21 vols. to date, New York, 1976–), v. 137.
7. 'Lord Howe and Lord George Germain: British Politics and the Winning of American Independence', *William & Mary Quarterly*, 3rd series, xxii (1965), 243.
8. *The Howe Brothers and the American Revolution* (New York, 1972), p.363.
9. 'Bamford's Diary', *Maryland Historical Magazine*, xxviii (1933), 16 and above, pp. 84, 85, 92–3.
10. '"To My Inexpressible Astonishment"', ed. Louis L. Tucker, *New-York Historical Society Quarterly*, xlviii (1964), 304.
11. Strachey to ..., [26 Aug. 1776,] (draft) Strachey MSS, DD/SH, C.1165, box 34, Somerset RO.
12. William L. Calderhead, 'British Naval Failure at Long Island', *New York History*, lvii (1976), 336, 338.
13. State Papers Ireland, SP 63/438, fo. 115, PRO.
14. Germain Papers, vol. 4, William L. Clements Library.
15. Entry of 19 Dec. 1776, Journal of Christopher French, Library of Congress.
16. Morris to John Hancock, 12 June 1777, (copy) Connolly MSS, Trinity College, Dublin.
17. Morris to Washington, 26 Jan. 1777, (draft) ibid.

18. Duff to the Earl of Fife, 22 April 1778, Duff of Braco Muniments, MS 2226/96/13, Aberdeen University Library.
19. Zetland (Dundas) MSS, ZNK X 2/1/46, North Yorkshire RO.
20. HMC, Fifteenth Report, Appendix, Pt. vi, *Carlisle MSS* (London, 1897), p.345.
21. *The Papers of George Mason*, ed. Robert A. Rutland (3 vols., Chapel Hill, N.C., 1970), i. 435.
22. *Diaries and Correspondence of James Harris*, ed. 3rd Earl of Malmsbury (4 vols., London, 1844), i. 400.
23. See, for instance, G.S. Graham, 'Considerations on the War of American Independence', *Bulletin of the Institute of Historical Research*, xxii (1949), 22–34; Michael Roberts, *Splendid Isolation 1763–1780* (Reading, 1970); Paul Langford, *Modern British Foreign Policy: The Eighteenth Century* (London, 1976), chapters 11 and 12.
24. Quoted in Jeremy Black, *A System of Ambition? British Foreign Policy 1660–1793* (London, 1991), p. 218.
25. Lord Lewisham to the Earl of Dartmouth, 29 Sept. 1775, HMC, *The Manuscripts of the Earl of Dartmouth* (3 vols., London, 1887–96), iii. 221.
26. *Franklin Papers*, xxiii. 196.
27. Ibid., xxv. 306.
28. Ibid., xxv. 727.
29. Lt. William Home to William Home, 19 June 1779, Logan Home of Edrom Muniments, GD 1/384/6/10, SRO.
30. *Sandwich Papers*, iii. 37.
31. Quoted in Jonathan R. Dull, *The French Navy and American Independence* (Princeton, N.J., 1975), p.133.
32. See, for example, Marshall Smelser, *The Winning of Independence* (New York, 1973), p.252; David French, *The British Way in Warfare* (London, 1990), p.76. Spanish scholars, of course, have emphasized their country's contribution for many years: see, for instance, Manuel Conrotte, *La intervención de España en la independencia de los Estados Unidos de la América del Norte* (Madrid, 1920); Francisco Morales Padrón, *Spanish Help in American Independence* (Madrid, 1952).
33. *John Jay: The Winning of the Peace*, ed. Richard B. Morris (New York, 1980), p.146.
34. *The Morning Chronicle and London Advertiser*, 27 Dec. 1780, quoted in David Starkey, 'British Privateering against the Dutch in the American Revolutionary War, 1780–1783', *Studies in British Privateering, Trading Enterprise and Seamen's Welfare*, ed. Stephen Fisher (Exeter, 1987), p.11.
35. Entry of 30 Dec. 1780, Diary and Account-book of Matthew Flinders, 1775–84, Lincolnshire Archives Office.
36. See H.M. Scott, *British Foreign Policy in the Age of the American Revolution* (Oxford, 1990), pp.307–9.
37. *Histoire de la participation de la France a l'etablissement des Etats-Unis d'Amerique*, ed. Henri Doniol (5 vols., Paris, 1886–99), v. 84.
38. Robinson Papers, L 30/14/307, Bedfordshire RO.
39. Rodney Papers, PRO 30/20/21/2, fo. 75.
40. Quoted in Jonathan R. Dull, *A Diplomatic History of the American Revolution* (New Haven, Conn., 1985), p.153.

10

The Legacy

For the British an important consequence of the war was great indebtedness. Borrowing, as we have seen, was vital to the funding of the British war effort. By 1783 the National Debt stood at a towering, unprecedented, and for many contemporaries a frightening, £232 million. To meet the interest charges alone required £9.5 million per year. As part of his financial reform programme, Chatham's son, William Pitt, who became prime minister at the end of 1783, set about cutting public expenditure. There were drastic reductions in the size of the army and navy, and there was a concerted attack on redundant civil offices: between 1784 and 1793 some 440 places in the revenue services were abolished. It was normal, of course, for post-war governments to behave in this way. Similar cuts were instituted by Henry Pelham after the War of Austrian Succession and by Grenville after the Seven Years' War. In real terms Pitt may indeed have achieved little more than Pelham or Grenville. But Pitt sought a much bigger overhaul than he was able to carry out – vested interests vitiated his efforts – and during his ministry there was, as John Brewer cautiously puts it, an 'extension of a reformist sensibility'.[1] In this respect Pitt was continuing the labours of Rockingham's second administration, which during its brief life put into practice many of the 'economical reforms' it had urged in opposition; though, ironically, it was Lord North who laid the foundations for Pitt's retrenchment campaign. To silence calls for greater financial stringency, in 1780 North had appointed commissioners for examining the public accounts. What was intended as a sop to appease the opposition merely intensified the clamour for reform. The detailed reports of the commissioners opened up to parliamentary and public scrutiny many unedifying aspects of the political system.

If North's diversionary tactic back-fired, it had important long-term repercussions. As successive ministers cut the number of offices and posts in public pay, one of the main props of eighteenth-century government was gradually weakened. It became more and more difficult for ministers to control Parliament through a body of loyal followers who had been given public office (or whose relatives had been made similarly dependent on public money) and who therefore were likely to vote as directed in the division lobbies. The erosion of this old pillar of power was halted and then put into reverse, of course, when the next war came along. Between 1793 and 1815 government patronage vastly expanded as the army, navy and revenue services grew. But with the close of the Napoleonic wars, and especially with the repeal of the income tax in 1816, the pressure to reduce public expenditure was stronger than ever. Over the next few years the number of public offices was severely cut. By the 1820s governments could no longer rely on a large corps of 'placemen' to support them in Parliament. Into the gaps created by the contraction of patronage seeped party. By the 1830s – perhaps before – governments were based more on party loyalty than on patronage. The growth of party itself cannot be directly attributed to the American war; the process was a long and interrupted one. But the decline of patronage, which was certainly a spur to the development of party, can be traced back to the commissioners for examining the public accounts and the reforms of the Rockingham and Pitt administrations.

The war was responsible for another political change of long-term significance. The monarchy became more popular. George III started his reign in 1760 with the advantage that he was perceived to be less German than his Hanoverian predecessors. Even so, he was suspected by a sizeable part of public and parliamentary opinion of harbouring 'despotic' ambitions. It was claimed that the young king wanted to restore monarchical power at the expense of ministers and Parliament; under the influence of his despised Scottish favourite Bute, George was said to be trying to undermine the constitutional settlement of 1689. Extra-parliamentary radicals, and some members of the parliamentary opposition, seem still to have believed this during the war years: like the American colonists they feared a plot to restore a Stuart-style tyranny. But by the general election of 1784 public opinion was willing to support George and his chosen minister against those politicians like Fox who wanted to curb monarchical influence. George's determination to sustain parliamentary supremacy over the colonies had helped to make him more popular, for in the American dispute he appeared to many Britons merely to be contending for the rights of the legislature, not seeking to increase his own authority. And George's stoical dignity in defeat added further to his reputation: he became a symbol of national determination. It was surely no coincidence that in the years following the American war 'God Save the King' started to supplant 'Rule Britannia' as the most popular patriotic song. The crown's position was strengthened yet more in the 1790s, when Britain's monarchical stability could readily be contrasted with France's republican anarchy; but the 'Apotheosis of George III', as

Linda Colley calls it,[2] started in the aftermath of the War of Independence.

George's new popularity can be associated with the emerging national consciousness which was explored in Chapter 8. It was not surprising, perhaps, that the war should further the development of national identity – threatening outsiders are a vital prerequisite to the creation of a sense of internal cohesion. Rather more curious is the way in which defeat apparently did little to diminish national pride. Indeed, the strength of the enemies ranged against Britain was seen as a cause for self-congratulation. John Mells, a customs officer in the Wainfleet district of Lincolnshire, noted in his day-book in May 1782: 'The Spirit of the British Lion is at last roused ... notwithstanding Great Britain has to contend with the House of Bourbon America and Holland'.[3] The vicar of Hanbury in Staffordshire was just as proud of Britain's performance against the odds: 'To future ages', he wrote in his parish register on 1 January 1783, 'it will appear to be an incredible Thing, what however, we know to be a Fact, that these Kingdoms sh[d]. maintain (as they have done) a glorious, but unequal, Conflict for several years, with the most formidable & unprovoked Confederacy that could be formed against them'.[4] It was as though defeat in these circumstances hardly counted as defeat at all.

Within a few years there were more solid reasons for regarding the American war as less than disastrous. The loss of the 13 colonies did not spell the end of the British empire. In the east, there was rapid expansion: the East India Company became more and more dominant in India, and by 1788 imports from Asia were worth £3.4 million. In the west, Canada grew with the influx of American loyalist refugees, and by 1791 there were sufficient Protestant settlers to justify the establishment of representative legislatures (though the Canada Act of that year carefully avoided giving the assemblies too much power over local executive officers). Even the West Indies recovered their profitability.[5] In 1780 and 1781 hurricanes and tropical storms caused extensive damage; but the war and its associated problems were probably still more devastating. The loss of islands to the French, the destruction of property, and the death of perhaps as many as 15,000 slaves – mainly due to the severe reduction in food supplies from the mainland North American colonies – inevitably diminished the staple–producing capacity of the region, as did the belated and inadequate attempts to cultivate more provisions on the islands themselves. Exports were further hit by the activities of enemy naval vessels and privateers. In 1772 the value of British imports from the Caribbean had stood at £3.4 million; in 1781 it dipped below £2 million. But after the war, despite British restrictions on trade with the new United

States, the islands bounced back. Jamaica, in particular, increased its sugar output; and between 1785 and 1790 the annual value of West Indian exports to Britain was consistently higher than for the years 1770–5.

Recovery at the international level came within a few years, too; though the signs were not at first propitious. A commercial treaty with France was signed in 1786, but it did little to improve relations between the old enemies. Indeed, fear of French ambitions, particularly in India, but also nearer home – in the United Provinces and even in Ireland – continued to drive British policy in this period. The aim of successive post-war foreign secretaries was to restore Britain's standing on the continent by negotiating an alliance with one or more of the European powers. But in the immediate aftermath of the American conflict they were no more successful than their predecessors had been during the war. Britain was perceived by potential partners as weakened by her defeat. Austria, Prussia and Russia declined to respond positively to British approaches. In 1787, however, the British were able to persuade the Prussians to invade the United Provinces to support the beleaguered Prince of Orange against the pro-French Patriot party. To deter the French from intervening, the Royal Navy was put in a state of readiness. The French – with domestic problems of their own – backed down. The British now had allies in the form of the Dutch and the Prussians, and French prestige had suffered an enormous blow. In 1790 France, convulsed by the beginnings of the Revolution, was again humiliated. A jurisdictional tussle between Louis XVI and the new National Assembly prevented the French navy coming to the assistance of Spain in her dispute with the British over Nootka Sound on the west coast of modern Canada. Unsupported by the French, the Spanish were compelled to retreat.

Alongside diplomatic recovery came economic recovery. The economic difficulties and dislocations of the American war were, for the most part, short-lived. Production of broadcloth in the West Riding totalled 4,845,000 yards in 1785, compared with only 3,224,000 yards in 1772. Narrowcloth output, which had stood at 2,378,000 yards in 1772, was up to 4,058,000 yards by 1787. Nor was the textile industry specially favoured. True, British re-exports remained below pre-1775 levels until 1792. This reflected, no doubt, a fall in tobacco imports from America (much of the tobacco had been re-exported to France before the war, now it could go straight there from the United States) and the collapse of the American market for re-exported Oriental goods (many of these were now imported directly by the United States). British-consumed imports, however, had topped the level of 1775 by 1784, and exports were back at their pre-war value by the last year of the conflict. Both continued to grow healthily until by 1792 imports were 26% higher than in 1772 and exports 40% higher. This rosy picture was the result of a variety of changes, one of which was the take-off of cotton textile production. But also important was the restoration of commercial contact with America. In 1784–6 the United States accounted for more than a fifth of all British exports. While Lord Shelburne's dream of Anglo-American free trade paving the way to some kind of political union was not to be realized, his hope that the United States would become a major trading partner was certainly fulfilled.

*

Recovery came more slowly to America. National and state debts were enormous. To service them taxes were higher than ever before (in Massachusetts they provoked open rebellion from farmers in the centre and west of the state). And while the tax burden was eating into American incomes, many of those incomes were themselves shrinking. We have seen that the war caused considerable physical damage to many parts of America. To replace buildings and equipment and to return farms to productivity took time and money. A South Carolinian planter noted in April 1783 that 'Since the War, the proceeds of our estates have been comparatively nothing'. 'It is really shocking', he continued, 'to see the infinite number of people who, before the War, possessed abundance and lived in plenty now reduced to penury and want.'[6] Three years later, Washington claimed that it was 'wonderful to see how soon the ravages of war are repaired',[7] but he may well have exaggerated: he was eager to demonstrate to a French correspondent the resilience of the new republic. The south, which saw much of the fighting and experienced a great deal of destruction, seems to have still been badly affected. Even in the early 1790s exports from the region were running at about half their pre-war value. And other areas must have had their share of the suffering. According to one estimate, living standards across the nation fell dramatically – per capita income plummetted by a staggering 46% between 1774 and 1790, a decline comparable to that in the Great Depression of 1929–33. If we can trust the figures in Samuel Blodget's *Economica: A Statistical Manual of the United States of America* (1806), then as late as 1805 wealth per capita remained 14% lower than in 1774. As the authors of a recent survey conclude: 'The colonists paid a high price for their freedom.'[8]

The price might well have seemed excessive when Americans considered the political condition of the new nation. Foreign statesmen routinely said that the former colonies were too weak to maintain their independence. There were predictions that sooner or later the Americans would be forced back into a subservience to Britain. In the aftermath of the war, the United States struggled to raise an army, and was unable to expel the British from posts in its own western lands. The root cause of this unimpressive performance was increasingly thought to be the Articles of Confederation. By reinstating the old imperial requisition system, with each local legislature providing its quota for common needs such as defence, the Articles left Congress enfeebled. Like British ministers in the Seven Years' War, Congress found that local responses to its requests were decidedly patchy. A despairing Washington wrote in August 1786 that 'requisitions are a perfect nihility where thirteen sovereign independent disunited States are in the habit of discussing and refusing compliance with them at their option'.[9] Washington had been arguing along the same lines since the beginning of the war. So had other Continental army officers. We have seen how Nathanael Greene condemned the selfishness of the states and prayed for a greater commitment to the national cause. In a sense, then, the Federal Constitution of 1787, which vested tax-raising powers in Congress and therefore strengthened central authority, owed its origins to the experiences of the struggle against the British – even if it took the problems of the post-war years to persuade many Americans that a new system was needed. The Federal Constitution did not instantly transform America's standing in the world, but by helping to overcome the nation's financial difficulties, and by providing funds for a proper military and naval force, it at least made foreigners believe that the United States was a viable entity. It was not to become a great power until its own industrial revolution radically changed the character of the American economy, and that was a long way ahead.

Meantime, the American economy reverted, more or less, to its pre-war structure. There were, to be sure, some important differences. American merchants and planters, no longer restricted by the Navigation Acts, could send their goods where they wanted, or at any rate to many more places than before the war. By 1790–2 some 16% of American exports went to northern Europe, notably to Germany, the Low Countries and Scandinavia. Trade with the British West Indies declined sharply, due to new British restrictions, but the other Caribbean islands now accounted for 24% of American exports. Internally, more Americans were gearing their economic activities to the requirements of the market. There were even signs of a growth in manufacturing output: in 1794, Tench Coxe, extolling the virtues of the new nation, boasted that 'The manufactures of the United States have increased very rapidly since the commencement of the revolutionary war, and particularly in the last five years.'[10]

But the similarities to the pre-war pattern remain striking. A large number of American exporters, even though free to choose with whom they traded, decided to renew contact with British merchants and middlemen. Besides sharing a common language, the British could offer credit and expertise that many Americans found attractive. In 1790–2 Britain and Ireland were still the biggest foreign markets for their goods, taking 31% of all American exports. Furthermore, vast quantities of British manufactures continued to be consumed by Americans – even after new import duties were imposed by the Federal government. In 1784 British exports to the United States were valued at £3,679,000; in 1792 at £4,271,000. Notwithstanding the claims of Coxe, many of the American industries that had sprung up during the war withered under this British commercial onslaught; even in the last years of the conflict Americans had been turning to higher-quality and lower-priced British models. Politically independent the Americans might have been, but their economic relationship with Britain remained, to all intents and purposes, colonial.

*

France, on the face of it, was a major beneficiary of the American war. Her main enemy, Britain, had been defeated and stripped of most of her North American colonies. France herself had made territorial gains in Africa and the West Indies. From 1784 the French West Indian islands were permitted to import supplies from the United States, enhancing still further their competitive advantage over the British. Nor did the loss of much of the provision trade seriously damage Bordeaux, which continued to grow with the increased output of the islands – particularly St. Domingue, where there was rapid expansion of sugar and coffee production in the 1780s. By 1789 Bordeaux's overseas trade was worth some 250 million *livres*, compared with a mere 40 millions in 1724. Yet, despite a robust post-war commerce, France, as we have just seen, was forced to give way to the British in the Dutch crisis of 1787, and two years later she was in the first throes of revolution.

Historians have long pondered the possibility of a connection between the American and French Revolutions.[11] Some have tried to establish a direct link between French involvement in the American war and the tumultuous events of 1789 and after. Did French officers serving in America bring revolutionary ideas back home? Lafayette, who went to America as a volunteer and appears to have been infected by the libertarian spirit prevalent in the United States, certainly seems to fit the bill. He was to become a prominent figure in the first stages of France's Revolution.

No doubt there were other officers similarly affected: comte de Ségur relished the chance to fight for liberty, and looked forward to seeing more of it in his own country. But a good many French officers appear to have been indifferent or amused by America – when the comte de Clermont-Crèvecoeur noted that the Americans 'all believe themselves equal',[12] he was not necessarily expressing approval – and at least a few, like the Swedish-born Fersen, were decidedly hostile to everything that they saw across the Atlantic. Fersen, far from absorbing American principles, became a counter-revolutionary in 1789.

What about the common soldiers in Rochambeau's expeditionary force? Forrest McDonald has argued that in America the ordinary French troops, most of whom were from peasant stock, came into contact for the first time with a society where free land ownership was the norm. Once back in France, and out of the army, they spread the message and in 1789 'formed the dynamic element in a movement which guaranteed the completion of the revolution by the destruction of economic feudalism'.[13] McDonald points to a correlation between the geographical origin of France's American war veterans and the areas affected by agrarian riots in the summer of 1789. But his evidence, to borrow the words of Claude Fohlen, 'is less convincing than it appears'.[14] The common French soldiers in America actually saw little of the country. Apart from on the march to Yorktown, they were confined to camp at Rhode Island and kept well away from local people. Even when they did come into contact with Americans, communication problems would surely have made it difficult for the French soldiers to gain a clear understanding of the local systems of land tenure. That many of the soldiers should come from the areas that experienced serious rural unrest in 1789 is hardly surprising, and proves nothing. These areas – Lorraine, Burgundy, Franche-Comté, Normandy, Guyenne and Gascony – were over-populated and poor in the eighteenth-century. The same economic imperatives that impelled struggling peasants to join the army may well have encouraged them to rise up against their landlords.

A more generally accepted connection has also been questioned. It has almost become an historical *cliché* that France was driven to near-bankruptcy by her role in the American war, and that her financial crisis set in train the events that led to the Revolution. Robert Harris has argued that this assessment owes too much to the self-serving claims of Calonne, the controller-general of finances, who in 1786 revealed to Louis XVI the dangerous state of his kingdom's debts, revenues and expenditures. Harris reckons that the war cost France only about half the sum usually thought, and that Necker's borrowing was not as ruinous as Calonne maintained. At the end of the war, Harris tells us, the French debt stood at

3,315.1 million *livres* (or about £143.5 million), and required about 165.4 million *livres* (or £7.16 million) per year in interest payments. This, he is at pains to point out, compares favourably with Britain's post-war debt.[15]

Harris is surely right to stress Calonne's determination to place the blame squarely on his rival Necker. Calonne spent lavishly between 1783 and 1786 on such projects as the new naval harbour at Cherbourg. During this period he borrowed only 20% less than Necker did in the war. But Harris's revisionism does not destroy the link between the American conflict and the Revolution. While Calonne's policies hardly helped, a great deal of damage had already been done. His post-war extravagance was designed to reassure creditors; he and they were already nervous, and with good reason. France's war expenditure was far from negligible – even if we adopt Harris's deflated estimate, it was about 1,006 million *livres* (about £46 million) – and it was financed in large measure by borrowing. Necker's loans, moreover, were expensive compared with Lord North's. The French debt in 1783 was only 62% of the British, yet French interest payments were 75% of the British. Given the fragile state of French finances in 1776, Turgot's prophesies about the dire consequences of French intervention were essentially sound. The war may not have been the only cause of the financial crisis of 1786, but without the war the crisis would surely not have occurred – and the crisis set France on the slippery slope to revolution.

*

Spain was not so dramatically affected – at least in the short term. Her prewar public expenditure of around 454 million *reales* per annum (about £12.5 million) climbed to over 700 million (more than £19 million) from 1779, and with South American income severely cut by the disruptions of the conflict, and domestic tax increases difficult to secure, much of this had to be borrowed. But the coming of the peace meant that accumulated South American treasure flooded into the Madrid treasury, allowing the newly-established national bank to start to repay the loans. Although Carlos III continued to borrow to fund an extravagant court, Spanish finances recovered from the war relatively quickly.[16]

But if Spain's American empire provided the wherewithal to keep the Bourbon regime afloat, the Spanish government realized that the empire itself was under threat as a result of the outcome of the war. Spanish fears about the independence of the British colonies in North America – 'such an event', Floridablanca was reported to have said in 1777, '... would be the worst example to other colonies'[17] – certainly seem to have encouraged Carlos's ministers to take pre-emptive action to protect the integrity of their master's empire. The introduction in 1778 of measures to open up the colonial trade can be seen in this light, though this initiative was influenced by domestic economic considerations as well. The spectre of the United States more clearly lay behind Carlos's attempt in 1787 to abolish distinctions between American Spaniards and European Spaniards. Henceforth, in theory at least, *criollos* (American Spaniards) would be as eligible as *peninsulares* for public office in the 'mother country'. If a common identity could be forged, then the empire would surely be immunized from the North American contagion. Informed by the same concerns were the grand schemes of federative monarchy put forward during and after the war. José de Abalos, a royal official in Venezuela, argued in 1781 that there was a 'vehement desire for independence'[18] among Spanish Americans and that this could best be countered by restructuring the empire along federal lines. Aranda is thought to have recommended in 1783 the creation of three Spanish American princedoms, and in 1786 he envisaged a revitalized empire in which a Spanish prince would rule the southern part of the American possessions, while the northern territories remained under direct control from Madrid.

Further federative programmes were considered during the next reign, but none was actually implemented. Even if one of them had been adopted, there is no reason to believe that the Spanish Americans would have been satisfied. In 1791 Juan Pablo Viscardo, a Peruvian jesuit, dismissed out of hand 'the ridiculous system of *union* and *equality*'.[19] The *criollos* wanted more control over what they now saw as their own countries, not parity within some reformed pan-Hispanic construct. But if Viscardo's writings suggest that Carlos and his ministers had failed to find the correct remedy, they confirm that Madrid's diagnosis of the problem was sound. The Spanish ministers were right to be anxious about the anti-colonial precedent set by the successful birth of the United States. Viscardo's impatience with Spanish rule was markedly increased by the successful revolt of the Britain's American possessions. 'The valor with which the English colonies ... have fought for their liberty ... covers our indolence with shame', he announced to his fellow Spanish Americans. 'We have yielded to them the palm with which they have been the first to crown the New World by their sovereign independence.'[20] It would take many years for the South Americans to follow suit – Argentina made a start in 1810, and it was not until 1824 that the Peruvians finally defeated the Spanish – and then the spark that set South America alight was the collapse of the Bourbon monarchy in Spain itself. But whatever the immediate cause of the South American pyrotechnics, the charges had been laid long before – and the American Revolution and War of Independence appear to have played a key role in the preparatory work.

*

The American conflict accelerated Dutch commercial decline; the already-tottering *VOC* never really recovered. The war also brought immediate and prolonged domestic turmoil for the United Provinces themselves. We have seen how bitter divisions between the *Stadhouder* and his opponents inhibited Dutch policy before the republic entered the conflict: the war made the rift still more acrimonious, and within a few years led to the fall of the Dutch *ancien régime*. As we saw in the last chapter, the British government, among its other miscalculations, imagined that one of the benefits of a successful war against the Dutch would be the discrediting of the anti-Orangists and the strengthening of the hand of the Anglophile Willem. The opposite happened. As nominal commander-in-chief of the Dutch forces the prince found himself blamed for the disappointments, defeats, and humiliations of the war. The reputation of the House of Orange was seriously undermined, and the demands for wide-ranging change became more vociferous. America, moreover, inspired the malcontents: the overthrow of British authority was portrayed as a re-run of the epic Dutch revolt against Spanish rule, and those who saw the Orange regime as responsible for Dutch decline could now borrow the language of their transaltantic friends. In September 1781 an anonymous pamphlet, *Aan Het Volk van Nederland (To the Netherlands People*) called for the creation of a citizen militia and an end to the Orangist usurpation of ancient rights. The author was Baron Joan Derk van der Capellen tot den Pol. In 1775 he had opposed the recall of the Scots brigade, arguing that he would 'prefer seeing Janissaries hired' to fight the Americans 'rather than the troops of a free State',[21] and over the next few years he was to be a leading figure in the emerging Patriot party which attempted to restore the provinces to their former glory by offering a heady mixture of the recovery of old liberties and the introduction of a more open and representative system of government.

Van der Capellen was to die in 1784, by which time the coaltion of anti-Orangist forces which he had helped to form during the war was already crumbling. Aristocratic and mercantile opponents of the prince – members of the so-called regent families – became alarmed by the more extreme demands of the Patriots, and moved towards an accommodation with the Orangists. The Patriots themselves were divided into two branches – the moderates, who wanted to reduce Willem's power and modernize Dutch institutions by opening up local oligarchies to wider participation, and the members of the radical and paramilitary 'Free Corps', who wanted to introduce more thorough-going democratic reforms. But neither the defection of the regents nor the divisions within the Patriot movement made life appreciably easier for Willem. In Holland, a Patriot stronghold, he was stripped of his authority and obliged to leave The Hague. In September 1786 Free Corps and soldiers loyal to the prince clashed in Gelderland; the following spring Orangist troops were put to flight at Vreeswijk, near Utrecht. To rescue the humiliated Willem, his brother-in-law, the king of Prussia, sent military assistance. The Patriots were quickly subdued, and the prince was restored to power behind the protective shield of 20,000 Prussian bayonets.

For the Orangist regime, however, this was no more than a stay of execution. The Patriots were not eradicated. Their clubs and societies revived with the outbreak of the Revolution in France. Their chance came in January 1795 with the arrival of the invading armies of the new French Republic. The *Stadhouder* fled and the United Provinces became the Batavian Republic. Under the new order the Dutch were given much more say in their own internal government, but this democratization was achieved only at the cost of an almost total loss of independence. Ironically, the Patriot dream of a Dutch renaissance led to a state that was a mere appendage of France, 'a manniken of the Quai d'Orsay'.[22]

*

So much for the fortunes of the different participants. But what was the legacy of the conflict in military terms? Did it change the nature of warfare? Technologically, the answer must be no. True, the submarine made its debut at New York in September 1776, when the *Turtle*, invented by Yale graduate David Bushnell, made an unsuccessful attack on Lord Howe's flagship. We might also note that the British army adopted a breech-loading rifle for the first time: the weapon designed by Captain Patrick Ferguson was tested at Woolwich in June 1776 and then issued to a corps of riflemen who served under Ferguson's command during the Pennsylvania campaign of 1777. But submarines made no real impact on the war – nor, indeed, on any war until the twentieth-century – and Ferguson's rifle failed to catch on; every regular British infantry unit but Ferguson's own corps continued to use the 'Brown Bess' musket introduced in 1717. Other wartime technological developments can also be passed over quickly: the 'carronade', a heavy gun fitted to Royal Navy ships for close-quarters combat, was of fleeting importance; the special gun-carriage designed by a Hanoverian artillery officer to enable the elevated defenders of Gibraltar to fire down on the besiegers was of purely local value; the copper-sheathing attached to the hulls of many British naval vessels merely extended a pre-war practice.

On a tactical level it could be argued that the conflict furthered the development of light infantry and inaugurated a move away from the rigid linear formations that characterized European warfare in the eighteenth century. We have seen how on St. Lucia in December 1778 the British light infantry dumbfounded their French assailants. The Hessian *Jägers* won an enviable reputation in America, and the lessons they learned ensured that the Hessian light troops won special praise during the early stages of the war against revolutionary France. Adam Ludwig Ochs, who had served in America, was particularly noted for his success in the Flanders campaign of 1793. Another *Jäger* veteran, August von Gneisenau, the famous military reformer, boosted the light infantry component in the Prussian army; while Johann Ewald published in 1790 an influential treatise that formed the basis for light infantry training in several countries.

One might note more generally the way in which the experience of the American war encouraged armies to adopt what could be called the light infantry ethos. The British foot regiments quickly adapted to local conditions. The tricorne hats, long coats, breeches and gaiters of 1775 had been much modified by 1777. Contemporary illustrations, particularly the drawings of the Italian artist Xavier Gatta, suggest that General Howe's troops in the Pennsylvania campaign were wearing caps or slouch hats, cut-down jackets to aid movement, and practical trousers or overalls. There is some evidence that Burgoyne's army adapted their uniforms in a similar fashion. These changes came about because the British were fighting in a different way: they were using looser and more fluid formations than was customary in Europe at the time – a flexibility forced on them by the wooded terrain of much of the American theatre. Some of them became, in time, skilled forest fighters. When in 1785 Cornwallis witnessed the manoeuvres of the Prussian army he could hardly believe his eyes: 'the worst General in England', he wrote, would not dream of ordering his men as the Prussians did; 'two long lines coming up within six yards of one another and firing until they had no ammunition left; nothing could be more ridiculous'.[23]

But the tactical advances of the war have to be kept in perspective. We should remember that light troops were not a product of the struggle in America. The Austrians, the French and the British had used light toops as skirmishers and scouts more than 30 years before, in the War of Austrian Succession. They were employed to an even greater extent in the Seven Years' War, both in Europe and America. In 1771 the British army had formally recognized their role by adding a light infantry company to each infantry battalion: indeed, one of General Howe's qualifications for the American command was that he had trained the assembled light infantry companies at Salisbury camp. Even the developmental argument – the American war enhanced the status of light infantrymen and light infantry methods – can be questioned. The Hessians, admittedly, were true torchbearers of the new military thinking: after its American experience the Hessian army placed much more emphasis on individual combat, sharp-shooting and speed (the standard infantry step was increased from 75 to 108 paces per minute). The British army, however, seems to have made less progress in this direction. True, British troops were able to perform creditably against the armies of the French Revolution from 1793 – a creditable performance almost certainly influenced by the experience of the American conflict. But not all of the lessons of the American war had been fully absorbed. Indeed, there was something of a backlash against American methods. Traditionalists complained that light troops and light infantry tactics had become disproportionately important. Colonel David Dundas, who had not served in the rebel colonies, maintained that 'light infantry, yagers, marksmen, riflemen, &c. &c.' were essentially peripheral to military operations. He lamented that they had become 'the principal feature of our army', and longed to see them restored to their role as 'an accessory to the battalion'.[24] Dundas and those like him may not have won a complete victory; but they did enough to retard the move towards a more mobile and flexible approach.

The most important military development of the American conflict was neither technological nor tactical, but social. The real innovation was the first appearance, on a significant scale, of a people's war. Of course, one can easily romanticize America's war effort by underplaying the extent of coercion, the inequality of sacrifice, and the role of Old World methods, money and military resources. But the participation of a large proportion of the free male population and the political commitment of many of those who fought for the United States certainly suggest that the Americans became a people at arms nearly two decades before the French. Less recognized is the extent of popular mobilization in Britain. During the American war the British army and the Royal Navy recruited more men than ever before. A higher proportion of Britons and Irishmen served as soldiers, sailors, militiamen or volunteers than in any previous war. For Britain, as well as America, this was an ideological struggle; a struggle that divided the political nation. Its political, economic and social impact was greater than is usually realized; it affected, in one way or another, the lives of countless thousands – even millions – of people in the British Isles. No-one would deny that the French Revolutionary War was still more dramatic in its impact; but for Britain the events of 1775–83 provided a lengthy preview of the tests and strains to come.

Chapter 10 Notes

1. John Brewer, *The Sinews of Power* (London, 1989), p.87.
2. 'The Apotheosis of George III: Loyalty, Royalty and the British Nation', *Past & Present*, no. 102 (1984), 94–129.
3. Misc. Don 681/1, Lincolnshire Archives Office.
4. D 1528/1/4, Staffordshire RO, quoted in Linda Colley, *Britons: Forging the Nation, 1707–1837* (New Haven, Conn., and London, 1992), p.144.
5. The case for continued growth and vitality is put by Seymour Drescher, *Econocide: British Slavery in the Era of Abolition* (Pittsburgh, 1977); a restatement of the old view that the American war marked the beginning of the end for the British West Indies can be found in S.H.H. Carrington, *The British West Indies during the American Revolution* (Dordrecht, 1988).
6. John Lloyd Letters, D1628, bundle 2, Gloucestershire RO.
7. *Writings of Washington*, xxviii. 500.
8. John J. McCusker and Russell R. Menard, *The Economy of British America, 1607–1789* (Chapel Hill, N.C., and London, 1991 edn.), p.374.
9. *Writings of Washington*, xxviii. 503.
10. *A View of the United States of America* (Philadelphia, 1794), p.431.
11. The classic work in this regard is R.R. Palmer, *The Age of the Democratic Revolution* (2 vols., Princeton, N.J., 1959–64). For a counterview, which argues for the uniqueness of each of the revolutions of this period, see Ian Christie, *Stress and Stability in Late Eighteenth-Century Britain* (Oxford, 1984), Chapter 1.
12. *American Campaigns*, i. 48.
13. 'The Relation of the French Peasant Veterans of the American Revolution to the Fall of Fuedalism in France, 1789–1792', *Agricultural History*, xxv (1951), 151–61.
14. 'The Impact of the American Revolution on France', *The Impact of the American Revolution Abroad*, ed. Richard B. Morris (Washington, D.C., 1976), p. 27.
15. 'French Finances and the American War, 1777–1783', *Journal of Modern History*, xlviii (1976), 233–58.
16. See Jacques A. Barbier and Herbert S. Klein, 'Revolutionary Wars and Public Finances: the Madrid Treasury, 1784–1807', *Journal of Economic History*, xli (1981), 315–39.
17. Quoted in John Lynch, *Bourbon Spain, 1700–1808* (Oxford, 1989), p.320.
18. Quoted in Mario Rodriguez, 'The Impact of the American Revolution on the Spanish- and Portuguese-Speaking World', *The Impact of the American Revolution Abroad*, p.103.
19. Quoted in Anthony Pagden, 'Identity Formation in Spanish America', *Colonial Identity in the Atlantic World, 1500–1800*, ed. Nicholas Canny and Anthony Pagden (Princeton, N.J., 1987), p.65.
20. Quoted in Rodriguez, p.115.
21. Capellen to Richard Price, 14 Dec. 1777, *The Correspondence of Richard Price*, ed. D.O. Thomas and Bernard Peach (2 vols. to date, Durham, N.C., and Cardiff, 1983–), i. 262.

22. Simon Schama, *Patriots and Liberators: Revolution in the Netherlands, 1780–1813* (London, 1991 edn.), p.2.
23. *Correspondence of Charles, First Marquis Cornwallis*, ed. Charles Ross (3 vols., London, 1859), i. 212.
24. David Dundas, *Principles of Military Movements, Chiefly Applied to Infantry* (London, 1795 edn.), p.12.

Bibliography

The sources available for a study of the American war are vast: it would take a lifetime (or longer) to read all the relevant manuscripts, published letters and journals, books and articles. Other researchers in the field might be interested to know which manuscript materials were consulted for this book, and for their benefit they are listed below. A sample of the published primary sources follows: it would consume far too much space to identify more than a tiny fraction of those available. Undergraduate students and general readers will perhaps be more interested in the 'Further Reading' section, which contains a selection of the secondary literature.

Primary sources

Manuscripts

Aberdeen University Library

Duff of Braco Muniments

Alnwick Castle, Northumberland

Percy Papers (now on deposit in the Northumberland Record Office)

American Museum in Britain, Bath

Peter Cornell MSS

American Philosophical Society, Philadelphia

Thomas Sullivan Journal

Archives Nationales, Paris

Archives de la Marine

Ardchattan Priory, Connell, Argyll

Campbell-Preston of Ardchattan Papers

Ballindalloch Castle, Banffshire

Macpherson-Grant of Ballindalloch Papers

Bedfordshire Record Office, Bedford

Polwarth Papers

Robinson Papers
Rugeley Papers

Berkshire Record Office, Reading

Abingdon St. Nicholas Parish Records
Neville and Aldworth Papers
Quarter Sessions Records
Thatcham Parish Records

Boston Public Library, Boston, Mass.

Miscellaneous Letters, Papers and Order-books
Percy Letters
Thomas Stanley Letters

The British Library, London

Althorp Papers
American Papers (Egerton MSS)
Auckland Papers
Benenden Letters
General Boyd's Gibraltar Log-book
Burgoyne Expedition Papers (Microfilm)
Journal of Alexander Chesney
Diario del Campo de Gibraltar
Diary of William Digby
Journal of Midshipman James Francis Grant
Haldimand Papers
Hardwicke Papers
Hastings Papers
Hutchinson and Oliver Papers (Egerton MSS)
John Johnson Letters
Liverpool Papers
Loudoun Papers
Macartney Papers
Mackenzie Papers
Martin Papers
Miscellaneous Papers (Microfilm)
Miscellaneous Papers relating to the Spanish Navy
Napier Papers
Nelson Papers
Newcastle Papers
Parker Papers (Microfilm)
Rainsford Papers
Siege of Gibraltar Journal, 1779
William Vassell Letter-books (Microfilm)

The Brotherton Library, University of Leeds

Jonathan Akroyd Account-books
John Jowitt Jnr.'s Debtor and Creditor Ledger
James Lister Day-book

Buckinghamshire Record Office, Aylesbury
Howard-Vyse Deposit

Calderdale District Archives, Halifax
Shibden Hall Muniments

Cambridgeshire Record Office, Cambridge
Huddleston of Sawston Papers
Quarter Sessions Records

Canna House, Isle of Canna, Scotland
Campbell of Inverneill Papers

Centre of South Asian Studies, Cambridge
Cromartie Papers (Photocopies)

Corporation of London Record Office
Sir John Langham's Charity Papers
Press Act papers
Quarter Sessions Records

Cumbria Record Office, Kendal
Letters of Robert Rauthmell

Devon Record Office, Exeter
Drake of Buckland Abbey Papers
Kennaway Papers
Quicke Papers
Miscellaneous Shipping Papers

Dr Williams's Library, London
Wodrow-Kenrick Correspondence

East Suffolk Record Office, Ipswich
Barrington Papers
Saumarez Papers

Essex Record Office, Chelmsford
Robert Greenwood Commonplace-book
Quarter Sessions Records
Round MSS

Glasgow University Library
Bannerman MSS

Gloucestershire Record Office, Gloucester
Blathwayt Papers
Ducie and Morton Muniments
Hale Papers

John Lloyd Letters
Rooke Papers

Hertfordshire Record Office, Hertford

Baker MSS
Ratcliffe MSS
Verulam MSS

The Hirsel, Berwickshire

Douglas-Home Muniments

Historical Society of Delaware, Wilmington

Dansey Papers

Historical Society of Pennsylvania, Philadelphia

Richard Bernard Diary
Chester County Miscellaneous Papers
Clifford Papers
Tench Coxe Papers
Dreer Collection: Daniel Weir Letter-book
Miscellaneous British Orderly-books
Pemberton Papers
Sir John Wrottesley Notebook

Hull University Library

Hotham Papers

India Office Library and Records, London

Bombay Letters Received
Diary of an Unidentified Officer
Home Miscellaneous Papers
Hughes MSS
Madras Letters Received
Munro Collection

Kent Archives Office, Maidstone

Amherst Papers
Twisden Family Papers

Library of Congress, Washington, D.C.

Elizabeth Jacquelin Ambler (Carrington) Papers
Joseph Ball Letter-book
Andrew Bell Papers
Robert Beverley Letter-book
Thomas J. Clay Papers
Colebrooke, Nesbitt, Colebrooke & Franks Papers
Stephen Collins & Son Collection
William Croghan Papers (Draper MSS)
Dunlop Family Papers

Peter Force Collection
Christopher French Journals
Nathanael Greene Papers
Hasbrouk Family Papers
Neil Jamieson Papers
Garret Minor Papers
Miscellaneous British Order-books
Miscellaneous MSS Collection
Robert Pleasants Letter-book
William Reynolds Letter-book
Abner Sanger Journal
Hezekiah Smith Papers
Benjamin Tallmadge Scrapbook
George Washington Papers

Lincolnshire Archives Office, Lincoln

Diary and Account-book of Matthew Flinders the Elder
Journal or Day-book of John Mells
Stubton Papers

Liverpool Record Office

Tarleton Papers

Massachusetts Historical Society, Boston

Belknap Papers
Coffin Family Papers
Caleb Davis Papers
Gardiner, Whipple and Allen Papers
Miscellaneous British Orderly-books
Miscellaneous MSS
J.M. Robbins Papers
Samuel P. Savage Papers

National Army Museum, London

Services of James Green
Laye Papers
Maitland Papers
Orderly-book, Fourteenth Foot, 1776

National Library of Scotland, Edinburgh

Halkett Papers
Fettercairn Papers
Houston Papers
John Macleod Letters
Maxwell, MacMurdo and Newhall Papers
Robertson-MacDonald Papers
Steuart Papers
Stuart-Stevenson Papers
William Wilson Papers

New-York Historical Society

Bancker Papers
Beekman Papers
Byvanck-Bleeker Papers
Duane Papers
Andrew Elliot Letters
William D. Faulkner Day- and Account-books
Neil Jamieson Day-book
Journal of John Miller (Reed Papers)
Miscellaneous British Orderly-books
Miscellaneous MSS
Charles Nicoll Papers
Frederick and Philip Rhinelander Papers
Stephen Skinner Letter-book

New York Public Library

Balch Papers
Bayard-Campbell-Pearsall Papers
Emmet Collection
Robert Henderson Day-book
Journal, Forty-seventh Foot (Photocopy)
John Kissam Papers
Nathaniel Lyttleton Savage Account-book

New York State Library, Albany

Evert Byvanck Papers
John Campbell Papers
Daniel Claus Papers
Andrew Elliot Papers
Israel Keith Papers
Miscellaneous British Orderly-books
Philip Skene Papers

Northumberland Record Office, Newcastle

Ridley (Blagdon) MSS

North Yorkshire Record Office, Northallerton

Wyvill MSS
Zetland (Dundas) MSS

Nottingham University Library

Mellish of Hodsock MSS
Newcastle of Clumber MSS

Pembrokeshire Record Office (Dyfed Archives), Haverfordwest

Daniel Gwynne Letters

The Public Record Office, London

Admiralty Papers

Audit Office Papers
Carleton Papers
Chatham Papers
Clerk of Assize Papers
Colonial Office Papers
Cornwallis Papers
Documents of Unknown Ownership (Richard Hope Letters)
Foreign Office Papers
High Court of Admiralty Papers
Hoare (Pitt) MSS
Home Office Papers
Rodney Papers
State Papers
Treasury Papers
War Office Papers

The Public Record Office of Northern Ireland, Belfast

Hart Papers

Queen's Lancashire Regiment Museum, Warrington

Plymouth Citadel Orderly-book, 1771–9

Rammerscales House, Dumfries

Bell-Macdonald Papers

Royal Artillery Institution, London

British Orderly-books, 1777–8
James Pattison Papers

The Royal Society, London

Blagden Papers

The Scottish Record Office, Edinburgh

Abercairny Muniments
Breadalbane Muniments
Broughton and Cally Muniments
Campbell of Barcaldine Muniments
Clerk of Penicuik Muniments
Cuninghame of Thorntoun Muniments
Dalguise Muniments
Dundas of Ochtertyre Muniments
Gilchrist of Ospisdale Muniments
Hamilton Dalrymple of North Berwick Muniments
Hay of Belton Muniments
Henderson of Fordell Muniments
Leith Hall Muniments
Leven and Melville Muniments
Lindsay of Dowhill Muniments
Logan Home of Edrom Muniments

Maclaine of Lochbuie Muniments
Maxtone Graham Muniments
Miscellaneous Bundles
Miscellaneous Gifts and Deposits
Murray of Lintrose Muniments
Murray of Polmaise MSS
Robertson of Kindeace Muniments
Seafield Muniments
Yule Collection

Sheffield Central Library

Rockingham MSS
Spencer Stanhope of Cannon Hall Muniments

Society of Friends Library, London

Miscellaneous MSS

Somerset Record Office, Taunton

Strachey MSS

Staffordshire Record Office, Stafford

Bagot Papers
Congreve Papers
Dartmouth MSS
Hanbury Parish Records
Paget Papers

Trinity College Library, Dublin

Connolly MSS

Warwickshire Record Office, Warwick

Quarter Sessions Records
Ward-Boughton-Leigh of Brownsover Collection

West Suffolk Record Office, Bury St. Edmunds

Grafton Papers

West Yorkshire Archive Service, Bradford

Spencer Stanhope Collection

Whitehaven Library, Cumbria

John Bragg Collection

William L. Clements Library, Ann Arbor, Michigan

Sir Henry Clinton Papers
Dering Family Papers
Thomas Gage Papers
Germain Papers
Journal of Capt. George Philip Hooke
Mackenzie Papers

Miscellaneous British Orderly-books
Miscellaneous MSS (Bound and Unbound)
Moncrieff Letter-book
Schoff Revolutionary War Collection
Simcoe Papers
George Wray Papers

William Salt Library, Stafford

Congreve Papers

Wiltshire Record Office, Trowbridge

Diary of Jonathan Adams
Savernake MSS
Wansey Papers

Worcestershire Record Office, Worcester

Quarter Sessions Records

Published Journals, Diaries, Letters and Memoirs

Leading Americans of the day have certainly not been neglected. Besides the authoritative editions of the papers of Washington, Franklin, Greene, Hamilton and Jefferson mentioned earlier in the 'List of Abbreviations', there are many others, such as the *Adams Family Correspondence*, ed. Lyman C. Butterfield (8 vols. to date, Cambridge, Mass., 1963–); *Papers of John Adams*, ed. Robert J. Taylor (8 vols. to date, Cambridge, Mass., 1977–); *The Papers of Josiah Bartlett*, ed. Frank Mevers (Hanover, N.H., 1979); *The Papers of Henry Laurens*, ed. Philip M. Hamer *et al.*, (12 vols. to date, Columbia, S.C., 1968–); *The Papers of William Livingston*, ed. Carl E. Prince (5 vols., Trenton and New Brunswick, N.J., 1979–88). Also very useful is the enormous collection of *Letters of Delegates to Congress*, ed. Paul H. Smith (21 vols. to date, New York, 1976–).

For the view from junior officers and the rank and file of the Continental army and militia there is also a rich literature, much of it the product of the assiduous editorial labours of members of American historical societies: see, for instance, the various diaries in Connecticut Historical Society, *Collections*, vii (1899) and 'Journal of Ebenezer Wild', Massachusetts Historical Society, *Proceedings*, 2nd series, vi (1890–1), 78–160. See also 'Journal of Sergeant William Young', *Pennsylvania Magazine of History and Biography*, viii (1884), 255–78; 'Sergeant John Smith's Diary of 1776', ed. Louise Rau, *Mississippi Valley Historical Review*, xx (1933–4), 247–70; 'Diary of Lieut. James McMichael', *Pennsylvania Archives*, 2nd series, xv (1890), 193–218. More recent additions to this genre include *Diary of a Common Soldier ... An Annotated Edition of the Military Journal of Jeremiah Greenman*, ed. Robert C. Bray and Paul E. Bushnell (DeKalb, Ill., 1978) and *Citizen Soldier: The Revolutionary War Journal of Joseph Bloomfield*, ed. Mark E. Lender and James Kirby Martin (Newark, N.J., 1982).

The 'Journal of Rev. Joshua Wingate Weeks', Essex Institute, *Historical Collections*, lii (1916), 1–16, 161–76, 197–208, 345–56, gives a flavour of the war from the viewpoint of a loyalist, as do the 'Letters of Robert Proud', *Pennsylvania Magazine of History and Biography*, xxxv (1910), 62–73. 'Silas Cooke – A Victim of the Revolution', ed. Susan S. Brayton, Rhode Island Historical Society, *Collections*, xxxi (1938), 108–21 paints a vivid picture of the misfortunes that could befall non-combatants. '"A Diary of Trifling Occurances": Philadelphia, 1776–1778, by Sarah Logan Fisher', ed. Nicholas B. Wainwright, *Pennsylvania Magazine of History and Biography*, lxxxii (1958), 411–65 and *The Diary of Elizabeth Drinker*, ed. Elaine Forman Crane (3 vols., Boston, 1991), are two sources that provide a female perspective.

The British side is almost as well served as the American. Sir John Fortescue's edition of *The Correspondence of George III* (6 vols., London, 1927–8) is not noted for its accuracy, but is still useful. Impeccably edited but perhaps less reliable for other reasons is Sir Henry Clinton's self-justificatory narrative *The American Rebellion*, ed. William B. Willcox (New Haven, Conn., 1954). We also have the *Correspondence of Charles, First Marquis Cornwallis*, ed. Charles Ross (3 vols., London, 1859). *The Private Papers of John, Earl of Sandwich*, ed. G.R. Barnes and J.H. Owen (4 vols., London, 1932–8) are invaluable, while *The American Journal of Ambrose Serle*, ed. Edward H. Tatum Jnr. (San Marino, Ca., 1940) gives us the views of Admiral Howe's secretary. Junior British officers produced a host of diaries and journals, many of which have been published. One of the best is Frederick Mackenzie's *Diary* (2 vols., Boston, 1930), which contains a wealth of detail and some illuminating insights into the mind of an intelligent and experienced officer. Also interesting is 'Lord Francis Napier's Journal of the Burgoyne Campaign', ed. S. Sydney Bradford, *Maryland Historical Magazine*, lvii (1962), 285–333. Published letters are still more plentiful; among the most accessible are *Letters from America*, ed. Eric Robson (Manchester, 1951) and *The Lost War: Letters from British Officers during the American Revolution*, ed. Marion Balderstone and David Syrett (New York, 1975), which covers naval as well as military operations, and Europe and the West Indies as well as North America. Hessian material includes *Revolution in America: Confidential Letters and Journals ... of Adjutant General Major Baurmeister*, trans. and ed. Bernard A. Uhlendorf (New Brunswick, N.J., 1957); Johann Ewald's *Diary*, trans. and ed. J.P. Tustin (New Haven, Conn., and London, 1979); and *A Hessian Diary of the American Revolution*, ed. Bruce E. Burgoyne (Norman, Okl., 1990), which comprises the daily record of an ordinary German soldier.

Life in wartime Britain can be sampled in two diaries edited by Basil Cozens-Hardy: *The Diary of Sylas Neville* (Oxford, 1950) and 'Mary Hardy's Diary', Norfolk Record Society, *Publications*, xxxvii (1968). Also helpful are Garland Cannon's edition of *The Letters of Sir William Jones* (2 vols. Oxford, 1970) and *A Selection of Thomas Twining's Letters*, ed. Ralph S. Walker (2 vols., Lampeter, 1990). For an Irish perspective see *Letters from an Ulster Land Agent*, ed. W.H. Crawford (Belfast, *1976)* and for the West Indies *The Diary of the Revd. William Jones*, ed. O.F. Christie (London, 1929).

A relatively small quantity of French material has been published. The collection of journals gathered by Howard C. Rice Jnr. and Anne S.K. Brown in *The American Campaigns of Rochambeau's Army* (2 vols., Princeton, N.J., and Providence, R.I., 1972) is excellent. Also useful are 'Meyronnet de Saint–Marc's Journal', trans. and ed. Roberta Leighton, *New–York Historical Society Quarterly*, xxxvi (1952), 255–87; 'Diary of a French Officer', *Magazine of American History*, vi (1880), 205–14, 293–308, 376–85, 441–52, vii (1881), 283–95; *The Revolutionary Journal of Baron*

Ludwig von Closen, trans. and ed. Evelyn M. Acomb (Chapel Hill, N.C., 1958); marquis de Chastellux's *Travels in North America*, trans. and ed. Howard C. Rice Jnr. (2 vols., Chapel Hill, N.C., 1963); *The Naval Campaigns of Count de Grasse*, trans. and ed. Amandus Johnson (Philadelphia, 1942). For the French in India Régine Pernoud's *La campagne des Indes: lettres inédites du Bailli de Suffren* (Mantes, 1941), is helpful. At the diplomatic level there is the weighty *Histoire de la participation de la France à l'etablissement des États-Unis d'Amerique*, ed. Henri Doniol (5 vols., Paris, 1886–99).

Spanish sources include 'Diary of Bernardo de Galvez', *Louisiana Historical Quarterly*, i (1917–18), 44–84; *Spain in the Mississippi Valley ... Pt. I: The Revolutionary Period, 1765–81*, trans. and ed. Lawrence Kinnaird (Washington, D.C., 1949); and *Spain Under the Bourbons*, trans. and ed. W.N. Hargreaves-Mawdsley (London, 1973). For the United Provinces there is very little in languages other than Dutch, though *Archives ou correspondance inédite de la maison d'Orange-Nassau*, ed. F.J.L. Krämer, 5th series (3 vols., Leiden, 1910–15), vol ii, contains some material in French.

Further Reading

The Background

The Anglo-American crisis has generated even more books and articles than the war. For the colonial period as a whole W.A. Speck's essay 'The International and Imperial Context' in *Colonial British America*, ed. Jack P. Greene and J.R. Pole (Baltimore, 1984) is a good starting point. Jack P. Greene, *Peripheries and Center* (Athens, Ga., 1986) is superb. For the breakdown of relations following the Seven Years' War Ian R. Christe and Benjamin W. Labaree, *Empire or Independence* (Oxford, 1976) covers both sides admirably, while R.W. Tucker and D.C. Hendrickson, *The Fall of the First British Empire* (Baltimore, 1982) is thought-provoking. P.D.G. Thomas has produced an authoritative trilogy from a British perspective: *British Politics and the Stamp Act Crisis* (Oxford, 1975); *The Townshend Duties Crisis* (Oxford, 1987); and *Tea Party to Independence* (Oxford, 1991). For American ideology there are Edmund S. and Helen Morgan, *The Stamp Act Crisis* (Chapel Hill, N.C., 1953); Bernard Bailyn, *The Ideological Origins of the American Revolution* (Cambridge, Mass., 1967); Gordon S. Wood, *The Creation of the American Republic* (Chapel Hill, N.C., 1969); and J.G.A. Pocock, *The Machiavellian Moment* (Princeton, N.J., 1975).

Edward Countryman, *The American Revolution* (London, 1987) concisely integrates political and social history, and reveals the extent to which social and economic themes have gripped the imagination of scholars in recent decades. See, for example, Jesse Lemisch, 'The American Revolution from the Bottom Up' in *Toward a New Past*, ed. Barton J. Bernstein (New York, 1968); Joseph Ernst, *Money and Politics in America* (Chapel Hill, N.C., 1973); Gary B. Nash, *The Urban Crucible* (Cambridge, Mass., 1979); T.H. Breen, *Tobacco Culture* (Princeton, N.J., 1985). Gordon S. Wood, *The Radicalism of the American Revolution* (New York, 1992) examines the transformation of American society before, during and after the war, while J.C.D. Clark, *The Language of Liberty 1660–1832* (Cambridge, 1994) stresses the importance of religious divisions in Britain and America.

The War: General

Among general histories of the conflict Piers Mackesy, *The War for America* (London, 1964), though focused on the strategy and problems of the British government, is outstanding. A new edition, with an introduction by John Shy, was published in 1993. Also valuable are Willard M. Wallace, *Appeal to Arms* (New York, 1951); Christopher Ward, *The War of the Revolution*, ed. John R. Alden (2 vols., New York, 1952); Eric Robson, *The American Revoltion in its Political and Military Aspects* (London, 1955); Howard H. Peckham, *The War for Independence* (Chicago, 1958); John R. Alden, *A History of the American Revolution* (London, 1969); Don Higginbotham, *The War of American Independence* (New York, 1971); Marshall Smelser, *The Winning of Independence* (New York, 1973); Robert Middlekauff, *The Glorious Cause* (New York, 1982); and Jeremy Black, *The War for America* (Stroud, 1991).

John Shy has brought together various of his stimulating essays in *A People Numerous and Armed* (New York, 1976), and Don Higginbotham has collected a number of his earlier writings in *War and Society in Revolutionary America* (Columbia, S.C., 1988). Professor Higginbotham has also edited a thought-provoking set of essays by different scholars: *Reconsiderations on the Revolutionary War* (Westport, Conn., 1978). Another important collection is *Arms and Independence*, ed. Ronald Hoffman and Peter J. Albert (Charlottesville, Va., 1984).

Military Operations

Particular battles, campaigns and theatres of operation are covered in the general studies above, and in a host of books and essays. Some of the older work now seems very dated - such as W.S. Stryker's *The Battles of Trenton and Princeton* (Boston and New York, 1898) and *The Battle of Monmouth*, ed. W.S. Myers (Princeton, N.J., 1927) – but there is a good deal of more recent literature. See, for instance, Russell F. Weighley, *The Partisan War: The South Carolina Campaign of 1780–1782* (Columbia, S.C., 1970); Jack M. Sosin, *The Revolutionary Frontier* (New York, 1967); A.T. Paterson, *The Other Armada: The Franco-Spanish Attempt to Invade Britain in 1779* (Manchester, 1960); T.H. McGuffie, *The Siege of Gibraltar* (London, 1965); S.P. Sen, *The French in India, 1763–1816* (Delhi, 1971). See also some of the essays in *The Revolutionary War in the South*, ed. W. Robert Higgins (Durham, N.C., 1979) and *Anglo-Spanish Confrontation on the Gulf Coast during the American Revolution*, ed. W.S. Coker and R. Rea (Pensacola, Fl., 1982).

The biographies of the commanders can of course be consulted with profit. On the American side we have John R. Alden, *George Washington* (Baton Rouge, La., 1984); J.T. Flexner, *George Washington in the American Revolution* (Boston, 1967); P.D. Nelson, *General Horatio Gates* (Baton Rouge, La., 1976); Theodore G. Thayer, *Nathanael Greene* (New York, 1960); Don R. Gerlach, *Proud Patriot: Philip Schuyler and the War of Independence* (Syracuse, N.Y., 1987); and *George Washington's Generals*, ed. George A. Billias (New York, 1964). Professor Billias has also edited a collection of essays on *George Washington's Opponents* (New York, 1969), and for more detailed coverage of the leading British military and naval commanders see John R. Alden, *General Gage in America* (Baton Rouge, La., 1948); William B. Willcox, *Portrait of a General: Sir Henry Clinton in the War of Independence* (New York, 1964); Franklin and Mary Wickwire, *Cornwallis: The*

American Adventure (Boston, 1970); and Richard J. Hargrove, *General John Burgoyne* (Newark, Del., 1983). Ira D. Gruber's *The Howe Brothers and the American Revolution* (New York, 1972) is a model of scholarship and historical writing. Also helpful are the biographies of two key British strategists: Gerald S. Brown, *The American Secretary: The Colonial Policy of Lord George Germain 1775–1778* (Ann Arbor, Mi., 1963) and N.A.M. Rodger, *The Insatiable Earl: A Life of John Montagu, 4th Earl of Sandwich* (London, 1993).

Armies, Navies and Money

For the British armed forces and their problems see Edward E. Curtis, *The Organization of the British Army in the American Revolution* (New Haven, Conn., 1926); J.A. Houlding, *Fit for Service: The Training of the British Army, 1715–1795* (Oxford, 1982); David Syrett, *The Royal Navy in American Waters* (Aldershot, 1989) and *Shipping and the American War* (London, 1970); R.A. Bowler, *Logistics and the Failure of the British Army in America* (Princeton, N.J., 1975); J.R. Western, *The English Militia in the Eighteenth Century* (London, 1965); Daniel Baugh, 'Why did Britain Lose Command of the Sea during the War for America?' in *The British Navy and the Use of Naval Power in the Eighteenth Century*, ed. Jeremy Black and Philip Woodfine (Leicester, 1988); and Nicholas Tracy, *Navies, Deterrance and American Independence* (Vancouver, 1988). Peter Mathias's 'The Finances of Freedom' in his *The Transformation of England* (London, 1979) compares British and American wartime public finances, while John Brewer, *The Sinews of Power* (London, 1989) makes British funding accessible – even exciting.

The allies and auxiliaries of the British can be studied in a number of monographs. Rodney Atwood, *The Hessians* (Cambridge, 1980) is the best modern account of the German troops, and Paul H. Smith's *Loyalists and Redcoats* (Chapel Hill, N.C., 1964) is splendid. See also Barbara Graymont, *The Iroquois and the American Revolution* (Syracuse, N.Y., 1972) and J.H. O'Donnell, *Southern Indians in the American Revolution* (Knoxville, Ky., 1973).

The American Continental army is considered from a variety of angles in Charles Royster's brilliant *A Revolutionary People at War* (Chapel Hill, N.C., 1979). Also valuable are Hugh F. Rankin, *The North Carolina Continentals* (Chapel Hill, N.C., 1971); K.K. Martins and M.E. Lender, *A Respectabale Army* (Arlington, Ill., 1982); R.K. Wright, *The Continental Army* (Washington, D.C., 1983); E.W. Carp, *To Starve the Army at Pleasure* (Chapel Hill, N.C., 1984). For the political struggles over leadership of the army see Jonathan G. Rossie, *The Politics of Command* (New York, 1975). Don Higginbotham has contributed a splendid essay on 'The American Militia: A Traditional Institution with Revolutionary Responsibilities' to *Reconsiderations on the Revolutionary War*. E. James Ferguson, *The Power of the Purse* (Chapel Hill, N.C., 1961) unravels the complex financial history of the early republic, while Robert A. Becker, *Revolution, Reform, and the Politics of American Taxation* (Baton Rouge, La., and London, 1980) is similarly helpful.

For the French see Lee Kennet, *The French Forces in* America (Wesport, Conn., 1977). Jonathan R. Dull, *The French Navy and American Independence* (Princeton, N.J., 1975) places naval history in diplomatic context. The difficulties facing the French navy are discussed in Michael Duffy's essay on 'The Foundations of British Naval Power' in the collection of pieces he has edited as *The Military Revolution and the State* (Exeter, 1980), by Daniel Baugh in the essay already mentioned, and by N.A.M. Rodger in 'The Continental Commitment in the Eighteenth Century' in *War, Strategy, and International Politics: Essays in Honour of Sir Michael Howard*, ed. Lawrence Freeman, Paul Hayes and Robert O'Neill (Oxford, 1992). J.F. Bosher, *French Finances, 1770–1795* (Cambridge, 1970) is useful, while Robert Harris paints a sympathetic picture of a key French finance minister in *Necker, Reform Statesman of the Ancien Regime* (Berkley, Ca., 1979). For Spain the best guide is John Lynch, *Bourbon Spain 1700–1808* (Oxford, 1989). For the Dutch Friedrich Edler, *The Dutch Republic and the American Revolution* (Baltimore, 1911) is still serviceable, but it needs to be supplemented by J.W. Schutte Nordholt, *The Dutch Republic and American Independence*, trans. Herbert H. Rowan (Chapel Hill, N.C., 1982) and Simon Schama's *Patriots and Liberators* (London, 1991 edn.)

Economy, Society and Politics

For the effect of the war on the American economy see Curtis P. Nettels, *The Emergence of a National Economy 1775–1815* (New York, 1962); Willard O. Mishoff, 'Business in Philadelphia during the British Occupation, 1777–1778', *Pennsylvania Magazine of History and Biography*, lxi (1937), 165–81; Elizabeth Cometti, 'Inflation in Revolutionary Maryland', *William & Mary Quarterly*, 3rd series, viii (1951), 228–34; Ann Bezanson, *Prices and Inflation during the American Revolution: Pennsylvania, 1770–1790* (Philadelphia, 1951); Julian Gwyn, 'The Impact of British Military Spending on the Colonial American Money Markets, 1760–1783', Canadian Historical Association, *Historical Papers / Communications historiques* (1980), 77–99. James A. Henretta has contributed a characteristically stimulating essay on 'The War for Independence and American Economic Development' to his *The Origins of American Capitalism* (Boston, 1991). Useful insights are also provided by Allan Kulikoff, *The Agrarian Origins of American Capitalism* (Charlottesville, Va., and London, 1992) and Jackson Turner Main, *The Sovereign States* (New York, 1973). Professor Main's work is valuable, too, on social change, on which he has also given us *The Social Structure of Revolutionary America* (Princeton, N.J., 1965). For the Philadelphia artisans we have Richard Ryerson, *The Revolution is now Begun: The Radical Committees of Philadelphia 1765–1776* (Philadelphia, 1978) and Steven Rosswurm, *Arms, Country and Class* (New Brunswick, N.J., 1987). On women see Mary Beth Norton, *Liberty's Daughters* (Boston, 1980); for Blacks Benjamin Quarles, *The Negro and the American Revolution* (Chapel Hill, N.C., 1961) and the essays in *Slavery and Freedom in the Age of the American Revolution*, ed. Ronald Hoffman and Ira Berlin (Charlottesville, Va., 1983). For local studies see Oscar T. Barck, *New York City during the War for Independence* (New York, 1931); Adrian C. Lieby, *The Revolutionary War in the Hackensack Valley* (New Brunswick, N.J., 1962); Donald Wallace White, *A Village at War: Chatham, New Jersey, and the American Revolution* (Rutherford, N.J., 1979); Richard Buel, *Dear Liberty: Connecticut's Mobilization for the Revolutionary War* (Middleton, Conn., 1980); Jerome J. Nadelhaft, *The Disorders of War: The Revolution in South Carolina* (Orono, Ma., 1981); John E. Selby, *The Revolution in Virginia* (Williamsburg, Va., 1988); Joseph S. Tiedemann, 'Patriots by Default: Queens County, New York, and the British Army, 1776–1783', *William & Mary Quarterly*, 3rd series, xliii (1986), 35–63.

On the British side the impact of war on the eighteenth-century economy was first explored by A.H. John in an article in *Economic History Review*, 2nd series, vii (1954–5), 329–44. As yet no systematic study dealing with the American war exists. One can learn a lot, however, from a variety of monographs and articles that touch upon the topic: Julian Hoppit's *Risk and Failure in English Business* (Cambridge, 1987), for instance, has a helpful section on war and includes material on the American conflict, and there are some nuggets in the sixth volume of *The Agrarian History of England and Wales*, ed. G.E. Mingay (Cambridge, 1989). See also T.M. Devine, 'Glasgow Merchants and the Collapse of the Tobacco Trade, 1775–1783', *Scottish Historical Review*, lii (1973), 50–74, and the figures in B.R. Mitchell, *British Historical Statistics* (Cambridge, 1988). For social aspects see J.M. Beattie, *Crime and the Courts in England, 1660–1800* (Oxford, 1986) and Linda Colley, *Britons: The Forging of the Nation 1707–1837* (New Haven, Conn., and London, 1992).

For 'high politics' – the politics of Westminster and the great country houses – see P.D.G. Thomas, *Lord North* (London, 1976); Ian Christie, *The End of North's Ministry* (London, 1958); and Frank O'Gorman, *The Rise of Party in England: The Rockingham Whigs 1760–82* (London, 1975). Bridging the gap between Westminster and popular politics is Herbert Butterfield's *George III, Lord North and the People, 1779–80* (London, 1949). For more on the Association movement see Ian Christie, *Wilkes, Wyvill and Reform* (London, 1962) and for the dissenters and the war James E. Bradley, *Popular Politics and the American Revolution in England* (Macon, Ga., 1986) and *Religion, Revolution and English Radicalism* (Cambridge, 1990). Irish developments can be traced in R.B. McDowell's magisterial *Ireland in the Age of Imperialism and Revolution, 1760–1801* (Oxford, 1979). For the military and political aspects of the Irish Volunteers see some of the articles in *Irish Sword*, xiii (1977–8). The impact of the war on local communities in Britain can be seen in such studies as Alastair Geddes, *Portsmouth during the Great French Wars, 1770–1800* (Portsmouth, 1970) and John Money, *Experience and Identity: Birmingham and the West Midlands, 1760–1800* (Manchester, 1977).

Diplomacy

The diplomacy of the war is covered in Richard B. Morris's monumental *The Peacemakers: The Great Powers and American Independence* (New York, 1965) and the first volume of Vincent Harlow's *The Founding of the Second British Empire* (London, 1952). Weldon A. Brown's *Empire or Independence: A Study in the Failure of Reconciliation, 1774–1783* (Baton Rouge, La., 1941) is still valuable. The standard account from the British perspective is now H.M. Scott, *British Foreign Policy in the Age of the American Revolution* (Oxford, 1990), but see also Isabel de Madriaga, *Britain, Russia and the Armed Neutrality of 1780* (London, 1962); Charles R. Ritcheson, 'The Earl of Shelburne and Peace with America, 1782–1783: Vision and Reality', *International History Review*, v (1983), 322–45; and Jeremy Black, *A System of Ambition? British Foreign Policy 1660–1793* (London, 1991). For the French we have Orville T. Murphy, *Charles Gravier, Comte de Vergennes: French Diplomacy in the Age of Revolution* (Albany, N.Y., 1982). The best overall study is Jonathan R. Dull, *A Diplomatic History of the American Revolution* (New Haven, Conn., and London, 1985).

Index